APPLIED ECOLOGY OF THE BLACK SEA

APPLIED ECOLOGY OF THE BLACK SEA

STRACHIMIR CHTEREV MAVRODIEV

Nova Science Publishers, Inc.
Commack, New York

Editorial Production: Susan Boriotti
Office Manager: Annette Hellinger
Graphics: Frank Grucci and John T'Lustachowski
Information Editor: Tatiana Shohov
Book Production: Donna Dennis, Patrick Davin, Christine Mathosian
    and Tammy Sauter
Circulation: Maryanne Schmidt
Marketing/Sales: Cathy DeGregory

Library of Congress Cataloging-in-Publication Data

Mavrodiev, Strachimir Chterev.
        Applied ecology of the Black Sea / Strachimir Chterev Mavrodiev.
                            p.   cm.
        Includes index.
        ISBN 1-56072-613-X
        1. Environmental management--Black Sea. 2. Black Sea.-- Environmental
conditions. 3. Marine ecology--Black Sea.. I. Title.
GE320.B48M38   1998   98-42433
333.91'64'0916389--dc21   CIP

Copyright © 1999 by Nova Science Publishers, Inc.
        6080 Jericho Turnpike, Suite 207
        Commack, New York 11725
        Tele. 516-499-3103   Fax 516-499-3146
        e-mail: Novascience@earthlink.net
        e-mail: Novascil@aol.com
        Web Site: http://www.nexusworld.com/nova

Printed in the United States of America

CONTENTS

PREFACE

This book is intended to increase the ecological education and responsibility of all sea users: Black Sea citizens, fishermen, holiday-makers, yacht-men, divers and other sea sportsmen; people dealing with tourism and health resort; workers and managers of enterprises and departments that pollute Black Sea; sailors and captains of ships navigating in Black Sea; dockers and chiefs of harbors; farmers and managers of agriculture and forestry; mariners, officers and admirals of Navy and frontier guard; members of the National Assembly and statesmen, who are always speaking about environment.

This book is a concise, revised and supplemented translation of the monograph "Applied Ecology of Sea Region - Black Sea", printed in the Publishing House of the Ukrainian Academy of Sciences "Scientific Thought", Kiev, September 1990, in accordance with the program of the Association of Black Sea Towns and Regions "Clean and Peaceful Black Sea".

The monograph has been written by scientists for who it is of importance the results of their work to find application in practice. The unsolved tasks are a challenge for us. Still more, we consider that so far our countries have not paid enough attention and had no clear strategy on the development and responsibility of applied ecology of sea. Insurance of the process of peaceful transition from totalitarism to democracy, development of free market economy, new time, all that gives us the chance as well as the duty to take upon our part of common responsibility.

The idea has generated in the end of the 70s in the Joint Institute for Nuclear Research - Dubna when a group of scientists, physicists and mathematicians, inspired by the first successes of "Nuclear Winter" model, pondered over the future of our civilization, the meaning and permissible limits of economic progress, necessity of scientifically motivated use of the limited resources of our planet, the role of science in making decisions and their realization.

As a first step for the vitality of these ideas, we made it ours aim to create the necessary scientific and management conditions preventing a future ecological catastrophe of Black Sea. Unfortunately, we acted rather slow. Only in 1986, a group was set up of scientists with close ideas about the future of our civilization and interests in Black Sea problems.

In 1987 we got the first financing by the Ministry of Science and Higher Education under Program Contract No. 435 together with the Base for Development and Implementation, Center of Physics of the Bulgarian Academy of Sciences.

In October 1988 we held our first Seminar "Pomorie 88", where:

a) protocol was signed for the establishment of an international scientific program group "Black Sea";

b) the program "Black Sea" was composed, employing the scientific conditions necessary for taking decisions on different managing, economical and technological levels, for prevention of Black Sea from ecological catastrophe;

c) a social offer has been accepted from the authorities of the town of Burgas for preparation of a "Report-Diagnosis on the State of the Burgas Bay and Recommendations for Mastering the Ecological Situation";

d) We formulated the idea for foundation of an Association of Black Sea Towns and Regions "Clean and Peaceful Black Sea" according to the principle "One sea one master".

Our group included scientists and specialists of academic and affiliated institutes in Russia, Ukraine, Georgia and Bulgaria, doing research of Black Sea: GEOHI (Moscow), Institute of Oceanology (Moscow), IHF (Moscow), Institute of Microbiology (Moscow), GIDROFIZIN (Sevastopol), INBUM (Sevastopol), Odessa Department of INBUM (Odessa), NPO "Gruzberegozashtita" Georgia (Tbilissi), INRNE B.A.S. (Sofia), IMH B.A.S. (Sofia), Faculty of Physics at SU "St. Kliment Ohridski" (Sofia), Program group No 435, BRV B.A.S., Sofia, Informational Center of Science and Implementation at the Ministry of Environment (Sofia), GOIN of GOSKOMGIDROMET (Moscow), Odessa Department of GOIN (Odessa), the Sevastopol Department of GOIN (Sevastopol), VNIRO of the Ministry of Fishing Industry (Moscow), Institute of Fishing Resources, "Fishing Industry" Firm (Burgas), Hydrographic service of CHF (Firm Sevastopol), Hydrographic service of BMF (Varna).

In the spring and winter of 1989 we created a data based program product "Black Sea" on the basis of "Argus" program product inserting the available historical data of Black Sea, collected information on the composition, regime and volume of wastewaters and other by-products thrown out in Black Sea, sources of polluting the Burgos Bay. We organized six ship's expeditions in the frames of the program

The regions of Burgos and Varna accepted and circulated the Appeal for foundation of an Association of Black Sea Towns and Regions "Clean and Peaceful Black Sea".

The Second Seminar "Pomorie 89" was held in 1989, 21 - 31 May, where:

a) data of Burgos Bay were processed and analyzed,
b) "Report-Diagnosis on the State of Burgos Bay and Recommendations for Mastering the Ecological Situation" was prepared;
c) Discussion on further advertising the idea for foundation of an Association was carried out.

The Report-Diagnosis was officially given to the authorities of Burgos on 29 of May 1989. On 14 of June 1989, the Executive Committee of the District Council of Burgos, taking into account the given recommendations in the Report, took decisions on mastering the ecological situation of Burgos Bay.

In spite of all efforts, only two practical results were obtained to our great regret:

a) much more attention was paid on the ecological ill-success of Burgos Bay and,
b) the zones for excretion of earth masses in Burgos and Varna bays were shifted in open sea near the main Black Sea stream.

As time showed, there was nobody to implement our recommendations, that is why the group decided to transform into a Firm research organization. In October 1989 we founded the Firm "ECOSY" (express ecological analysis, diagnosis and prognosis of sea regions for taking nature preserving decisions accompanying industrial activity, import-export, etc.).

We believed that this type of organization should have to allow us to overcome easier the scientific, topical, departmental, regional, party and national feudalism on the one hand, and to make own money and self-finance our work on the other hand.

The analysis of the situation showed that in spite of publicity still people are not well informed about the real state of Black Sea ecosystem, its near future and necessary measures.

That generated the idea of a monograph that would summarize all collected data on Black Sea.

At the same time we continued the investigations to get a digital - evaluation of the different sources that contribute to the pollution of Black Sea in the starting ecological catastrophe; we made every effort to realize our recommendations for Burgos Bay; we promoted further the organization of an Association of the Black Sea Towns.

In the Seminar "Pomorie 90"- "Applied Ecology of Black Sea", held from 21 May to 3 June 1990, 77 scientists were present as well as representatives of the local state authorities and industry, physicians and journalists. Among the scientists were 29 Ph.D., 7 D.Sc. and 4 directors of Institutes. Although the computer technique was insufficient, the manuscript of the monograph "Applied Ecology of the Sea Regions - Black Sea" was edited and printed in Russian in 400 pages, including tables, graphs, maps and drawings. The author's staff consisted of 103 scientists of the above mentioned institutes. The manuscript was prepared for print in the Publishing Department of the Sea Hydrophysical Institute, Sevastopol.

The initial translation and revision of the monograph has been done by senior res.Ph.D. Venelin Velev (IO B.A.S., Varna), res. Rumen Russev (INRNE B.A.S., Sofia) and myself. Additional editorial work has been done together with Rumen Russev. Significant technical and editorial assistance has been done by res. Elka Zlateva (INRNE B.A.S., Sofia) and students Rumyana Tsoneva and Kiril Russev from the Faculty of Physics at Sofia University.

This monograph will serve the reader by supplying the necessary references.

The program group owes the completion of its work to the benevolence of thousands of people: fishermen and sailors, directors of scientific and economic organizations, admirals and officers of Navy and frontier guards of Bulgaria and USSR, scientists and members of boards of the Bulgarian, Georgian, Russian and Ukrainian Academies of Science, local authorities, journalists. Special mention should be made of Academician Valeri Barsukov, of Nedelcho Pandev, of Professors Matei Mateev and Alexander Asenov, Stoicho Panchev, of captains Anton Yankulov, Ignat Tenchev, Kostadin Stoichev, of Mr. Svilenov (Rila), Mr. Dionisiev (Mladost), Mr. Petrov (Vinprom, Pomorie), of journalists Panayot Manolov and Lyudmula Sokolovskay, and of public worker Alla Shevchuk. Thank, also, to some scientific and state managers that suppressed their reluctance to allow something non-standard.

Posthumously, we express our deep gratitude to Professor Anatoliy Simonov, whose tragic death in the summer of 1989 was a great loss for us. Without his encyclopedic knowledge and universally recognized authority, we should have hardly had the courage to take upon this immense task. His sense of responsibility and duty, his enormous capacity of work are as a model for us. May his memory live for ever!

My friend Prof.Dr. Vachtang Garsevanichvily proposed to Dr. Frank Columbus from Nova Science Publishing Company to publish this book. So, at last our efforts to inform the people for the situation on the Black Sea ecosystem, are finished. Our great thanks to them!

Asst. Prof., Dr. Strachimir Chterev Mavrodiev
28 of October, 1990
Sofia, Pomorie

List of the editorial staff and authors of monograph "Applied Ecology of Sea Regions- Black Sea", Naukova Dumka, Kiev, 1990:

Editorial staff: S. G. Boguslavsky, V. Velev, G. Gergov, S.A. Efremov, A.P. Zhilyaev, I. Ê. Ivanchenko (secretary), V. P. Êeondjyan (editor-in-chief), R.D. Kosyan, A. Ì. Kudin (vice-editor), S. Cht. Mavrodiev, V.I. Mikhailov.

M. Ovsienko, I.M. Ovchinnikov, A. G. Pavlova, A.I. Simonov, YU.I. Skurlatov, YU.V. Terehin (vice-editor), Ì.V. Flint, M.S. Tsitskishvili;

AUTHORS: E.N. Altman, Ph.D., A.A. Bezborodov, Ph.D., YU.I. Bogatova, D.Sc., S. G. Boguslavsky, D.Sc., A.M. Bronfman, D.Sc., Z.P. Burlakova, Ph.D., B.S.Veleva, Ph.D. V.Velev, Ph.D., O.Å.Vinogradov, Corresponding member D.Sc., L. A. Vinogradova, Ph.D., L.V.Vorobyova, Ph.D., G.P. Garkavaya, Ph.D., G. Gergov, Ph.D., I.F. Ertman, Ph.D., Z.A. Golubeva, G.A. Grishin, Ph.D., M.B. Gulin, S.B. Gulin, A. Damyanova, Ph.D., YU.Ì. Denga, SH.V. Djaoshvili, D.Sc., S.A. Dozenko, V.N. Egorov, D.Sc., V.A. Ememelyanov, Ph.D., V.N. Eremeev, D.Sc., L.V. Eremeeva, Ph.D., S.A. Efremov, Ph.D., N.V. Zherko, Ph.D., A.P. Zilyaev, Ph.D., V.A. Zhorov, Ph.D., YU.P. Zaitsev, Corresponding member Ph.D., O.I. Zilberstein, Ph.D., I.K. Ivanshtenko, Ph.D., M.I. Kabanov, Ph.D., V.I. Kalatsky, D.Sc., S K• rapeev, V.S. Karpov, Ph.D., T.G. Kasich, V.P. Keondjyan, D.Sc., A.G. Kiknadze, D.Sc., R.D. Kosyan, Ph.D., V.G. Krivosheya, Ph.D., A.M. Kudin, Ph.D., V.S. Kuzhinovski, I.I.Kulakova, Yu.D.Kulev, N.V.Kucheruk, Ph.D., G.E.Lazorenko, Ph.D., S.P. Levikov, Ph.D., O.I. Leonenko, S.Cht.Mavrodiev, Ph.D., O.V.Maksimova, V.I.Medinets, Ph.D., L.Minev, Ph.D., V.I.Mihailov, Ph.D., I.N. Mitskevich, Ph.D., A.I. Nesterov, D.Sc., D.A. Nesterova, Ph.D., L.E. Nizhegorodova, Ph.D., G.G.Nikolaeva, Ph.D., S.N.Ovsienko, Ph.D., I.M. Ovchinnikov, D.Sc., I.G.Orlova, Ph.D., G.G.Polikarpov, Acad.D.Sc., L.N. Polishchuk, Ph.D., V. Popov, R. Rusev, N.I. Ryasintseva, Ph.D., P.T. Savin, Ph.D., A.F. Sazhin, Ph.D., Prof. V.V. Sapozhnikov, A.I. Simonov, D.Sc., I.A. Sinegub, Yu.I. Skurlatov, D.Sc., L.A. Stefantsev, Ph.D., I.N. Suhanova, Ph.D., V.I. Sychev, Ph.D., N.N. Tereshchenko, Ph.D., YU.V. Terehin, Ph.D., V.I. Timoshchuk, Ph.D., V.B. Titov, Ph.D., A. Tomov, A.P. Filipov, M.V. Flint, Ph.D., E.P. Hlebnikov, Ph.D., M.S. Tsitskishvili, Ph.D., V.G. Tsytsugina, T.V. Chudinovskih, Ph.D., E.V. Shtamm, Ph.D., E.A. Shushkina, Ph.D.

MOTIVES

Present civilization faces a serious choice. Nuclear winter is not the only way to self-destruction. The other one is the uncontrolled and non-scientific usage of planet's resources in the name of "progress". The rates and technologies of development of energetic, industry (especially chemical industry), agriculture and transport in the last fifteen years will bring to bad quality of air and drinking water in the next several decades. Warming up of planet will wash away the lines of seasons.

History teaches that the achievements of science sometimes find inhuman application. Let us only remind the Nobel invention, the tragedy of Hiroshima and Nagasaki.

The power of today's technique is already comparable to the natural processes of our planet. For this reason, the scientifically groundless interference or impact on the processes of nature in latter years leads to unforeseen and, as a rule, fatal effects: the global warming, the ozone hole and etc..

To continue the existence of our civilization, we should learn how to approach nature and exploit its resources on scientific grounds, along with the new attitude in the interstate relations. We should learn how to co-ordinate the rates and quality of material progress with the possibilities of planet. This is an exceptionally difficult scientific task, as almost recently it became possible for ecology and environment science to look for and find solutions with quantitative evaluations on the movement of different elements in nature, the connection among different communities, different geographic regions, forecasting and management of energy and resource industry, due to the impressive development of mathematics, physics, biology and powerful computerization of science.

Over than seventy years ago the Russian scientist Academician V.I. Vernadsky foresaw the contemporary ethical, ecological, energy and resource problems of our civilization. He outlined their solution in a qualitative way on the basis of a joint study of biosphere and the discovery that mankind turns into a geological power. However, the knowledge of nature law by scientists still does not mean its correct usage and application.

WHY BLACK SEA?

As it is known Black Sea is a closed sea, a water gathering basin of East and Middle Europe - a region with well developed energy, industry, agriculture, transport and tourism. In the end, all wastes of human activities fall into it. Of course, a part of the substances disintegrate on their way to sea. After 1970, their volume increased to such an extent that the impact on the Black Sea ecosystems began to be noticed considerably. Also, we have not to forget the undercurrent from the Sea of Marmara through Bosporus and the rainfalls driven by the Atlantic cyclones and the north-easterly wind that contribute to the pollution. The ecological breakdowns in the north-west part of sea, periodic blossoming of plankton, change in quality and quantity of fish catch, decrease of water transparency, appearance of a new settler Mnemiopsis leidyi, all that illustrates qualitatively this fatal influence on the Black Sea ecosystems.

The quantitative estimation of these phenomena, their interconnection and components of pollution, is an enormous field of interest for mathematicians, physicists, chemists, biologists and engineers of fundamental and applied research.

ULTIMATE GOAL

The stages of research should include:

1) Creation of a data bank containing the huge volume of information about sea and its ecosystems collected up to now and periodically enriched by,

2) A system for physical, chemical and biological monitoring, as well as,

3) Creation of a complex model of Black Sea ecosystems by recording their hierarchy and thus reliably describing the sea phenomena as well as their prediction.

It is obvious that the ultimate goal may be formulated on such basis only,

4) To "rule" scientifically the Black Sea ecosystem on terms of agreement among the interested countries.

There were scientists who call the pessimistic prediction for the inevitable ecological breakdown of Black Sea "ecological hysterics". Others believed that the self-cleaning capacity of sea would develop so that the ecosystems will take in the increasing pollution without much difficulty, as at the same time the quality of water and air will remain acceptable for people.

We know how to KNOW and NOT TO BELIEVE.

All Black Sea countries possess rather powerful science institutions and possibilities for sea investigations. However, all these costly efforts will remain, as until now, almost inapplicable and meaningless without functional co-ordination and a joint unified system for collection and processing of data, information about the results from investigations and predictions given to every citizen of Black Sea region, control on the measures of sea protection and publicity.

It is natural for the citizen of Black Sea towns and other sea users to be most interested in its cleanliness and take good care!

There is a saying:

"Help yourself alone, so that God will help you".

1

BRIEF GEOLOGICAL HISTORY OF BLACK SEA BASIN

Present Black Sea is heir of a series of ancient basins. Its evolution, particularly in its early stages of existence, different contours and dimensions, is not always easily traced in "the geology calendar". We shall point out only its last era - neozoic. Before its beginning, the ocean Tetis had stretched in the west and far in the south-east between the African continent and the Eurasian platform. 30-40 million years ago it divided in two - Tetis and para-Tetis (in the north) as several deep-water hollows had formed in their frames. About 6 million years ago they broke into smaller basins, separated by hills and mountain ranges. Gradually, most of these shallow basins filled up with depositions and their waters withdrew. Only the slightly salty Chaudin and Bakin basins remained - predecessors of present Black and Caspian Sea. At that time (about 600 thousand years ago), they had occupied its today geographic position.

Today's deep-water profile of Black Sea bottom had not been created "at once". To its long history testifies the bottom configuration of the ancient basin, whose bottoms are buried under several complexes of rocks and still untightened depositions.

The beginning of the Alpine orogeny formation in the Mediterranean (then - Tetis) had begun in Oligocene - Miocene, i.e. 30-10 million years ago. The ancient ocean Tetis had reached its maximum dimensions shortly before that and pressed by the raising mountain ranges had begun to yield. A vast peripheral sea had formed, oriented west-east (para-Tetis) in the present places of East Bulgaria, Crimea, Caucasus, having a connection with ocean in the region of Iran, Iraq and Syria.

The next clear seismic boundary crosses the upper part of Eocene deposits. Two huge "lens" of sediments are closed between it and the lowest reflecting boundary, deposited in both Black Sea depressions. The depositions over this boundary had been heaped mainly during Oligocene and the beginning of Miocene. Generally, they are limeless clays with sand and alevrit insertions. This deposition complex contains beds of oil and gas in many places in the north of Caucasus.

Miocene sediments, discovered during the soundings of ship "Glomar Challenger", are dark to black color fine-grained and clay sands. At that time, there had existed a comparatively shallow aquatic basin in the region of Bosporus that had deepened in the north. After its withdrawal from northeast Bulgaria, the Miocene sea had left behind horizons with biogenic limes (cliff coast from Cape Kaliakra to Tyulenovo); clay-marl deposits are spread mainly in the south coastline.

In the end of Miocene, more or less isolated basins had differentiated in the frames of east para-Tetis, situated in non-deep depressions among hilly and low mountain regions: Panonian, Dacca, Aegean, Caspian. The conditions of climate were favorable (the temperatures in winter were above +5°C, the rainfalls were sufficient), the land and seas were populated with rich flora and fauna.

In the end of the stage (6-5.5 million years ago) important geographic and climatic changes took place. The so called "Mesina crisis" had started playing in the Mediterranean: the sea separated from the Atlantic Ocean and sharply decreased its level. That was accompanied by the domination of dry, continental up to desert climate. The separated aquatic basins increased saltiness of their waters, decreased in volume and deposited great amounts of gypsum and rock salt.

5.5 million years ago the Pontian basin was replaced by Kimerian Sea. It had existed in conditions of subtropics climate (the average January temperature about +18 °C) and considerable humidity. However, after 2 million years a gradual cooling began as well as aridity of climate. The first glaciers were generated in the already high peaks of the Alps. Correspondingly, considerable changes took place in the flora and fauna.

About 2.9 million years ago in the end of Pliocene, the slightly salty Kuyalnik sea existed in the basin region. It had been succeeded by the Gurian basin in an interval of 1.6-1.1 million years and became still cooler. The tropic ferns disappeared in the exuberant forest massifs. The last of the basins is called Chaudin and existed about 800-500 thousand years ago. It has had the same configuration as Black Sea, but its water was from slightly salty to fresh. It distinguished by comparatively low water stay, due to regression and wide coastal flats. Basin and land were inhabited by numerous organisms. Climate was warmer than the present one (the average January temperature had not been lower than +5° C) and the rainfalls were considerable. Thick deciduous forests surrounded the Chaudin sea-lake thick forests changing into coniferous massifs in the mountain regions. Chaudin basin has had a connection with the Caspian sea, and in the end of the period with the Mediterranean. Approximately at that time the man had appeared in the Balkan Peninsula.

The history of the basin during the Quaternary (anthropogenic period) represents a series of warmings and coolings. Transgressions had taken turns with regressions, accompanied by corresponding climatic fluctuations. The cool episodes were best marked by the appearance and expansion (the movement) of mountain glaciers and the glacier cover in North Europe. These Glacial periods were four in the Alps and separated by interglacial epochs with warmer climate. At warming the rivers had become deep, carrying huge masses of fresh water toward lakes and sea basin. As a result, the water in them had lowered its mineralization and the basin advanced toward land, extending its boundaries (transgression). During the next cooling the rivers decreased their flow and, in consequence, the aquatic basin shortened their perimeter, i.e. regression took place.

The last interglacial warming, called Wurm in the region of the Alps, had taken place about 120 thousand years. It caused a great glacial eutectic transgression of ocean. Gradually it had also included the Mediterranean reaching to Black Sea. That had been the time of Karanga and the ancient basin is called "Karangan". With fluctuations and restraints Karangan sea had begun to raise its level exceeding even today's level. Fauna in the sea was salt-water, of Mediterranean type and this points to the existing connection between the two basins, which practically have had equal saltiness (about 30 parts of salt in 1000 parts of water).

The last global cooling of climate (Wurm) marked its maximum about 190 thousand years ago. During the greater part of late Wurm the Black sea basin shortened its contour. A great part of the basin shallow section transformed into a low levelled land, cut through river valleys and firths appeared in proximity to sea. The connection with the Mediterranean, realized in the region of the Straits and the Sea of Marmara, was cut off. For this reason, in the time of new Euksin (about 23 thousand years ago) the waters in the basin had refreshed in fact.

The new Euksin basin level had begun to rise hardly 12-14 thousand years ago, which might due to river flows and atmosphere waters (melting glaciers) or to global transgression. The bed of Bosporus threshold reached the roots and steady rocks at depth of 70 ì, and that had prevented mixing of waters in the beginning of the stage.

Some scientists call that 40 thousand years period a time of cosmic catastrophes; when maybe Earth obtained the Moon, the terrestrial axis had changed its direction, the magnetic field had also changed, Atlantis had disappeared.

About 8 thousand years ago the level had become to raise as the Bosporus stream started its action throwing over salt waters from the Mediterranean Sea into the deep water valley of Black Sea.

Some scientists connect this rise of level with the Biblical legend about the world Flood. The recent investigations of Greek scientists show a possible connection between the opening of Bosporus and Dardanelles and the Flood with the eruption of Santarin volcano.

Finding of shells in the region of Crimea that are typical for Baltic Sea and Arctic Ocean is accepted as a proof for the terrestrial axis shift.

A complex investigation of the region of Strandzha and Sakar mountains may give an answer to this exciting riddle. It must include investigations of sea sediments in the shelf, biological investigations of excavated seed, particularly the relict species, complex and archeological investigations, precise isotope dating of samples and restoration of the near stellar encirclement (up to 20- 40 thousand light years) of our Solar system.

This transgression called New Black Sea had reached its maximum range about 5 – 4.5 thousand years ago and the level of that time sea had exceeded the present level with 2 – 2.5 ì. About this period the water mass in the whole basin had become salty almost to the present extent for example, and a dense physico-chemical boundary had appeared in the uniform water mass (pinoclin and chaloclin). Fresh water had still remained only in firths.

The New Black Sea transgression had been followed by the Fanagorian regression with eustatic minimum achieved 500-300 years ago. This regression has shaped the present contour of Black Sea basin.

How long may live on Black Sea? The long geological history and the present reality, passing before the eyes of only one generation, show that most of the aquatic basins, in spite of their scale, early or late fill with depositions: gravel, sands and slime. It was already mentioned that Black Sea inherited the ancient ocean Tetis that gradually disintegrated and shortened. In the present tectonic stage, the continental slates of Africa and Europe, separated by the Mediterranean basin, gradually draw close almost at 1 cm per year. The Arabian and Eurasian slates also draw close, at that Asia Minor moves in the west. After about 50 million years the Black and Caspian Seas will not exist as sizeable basins. Much earlier, only after 6 million years, the Adriatic Sea will disappear because the Apennines and Balkan Peninsulas converge and will merge. The Mediterranean Sea as a whole will shorten its area and disintegrate into separate basins.

Other factors have to be also recorded in a "long-term prognosis". For example, if the present accelerated rates of deposition heaping in Black Sea remain, its volume will fill up for 2-3 million years. On the other hand, the global warming up of climate, caused by the "green-house effect" in atmosphere, where the content of carbon dioxide, methane and water vapours increase, contributes to the rise of the World Ocean and Black Sea level. Now this rise is estimated at about 1.5 mm per year for Black Sea in the last 15 years.

2

SOME OF THE RESOURCES OF BLACK SEA REGION

On 9 January 1987 Bulgaria declared its right of an exceptional economy zone in the basin of Black Sea. The precise boundaries of this zone are still not co-ordinated with our neighbours, but obviously, this will happen soon. This new space reveals new perspectives and problems for our country. There are different kinds of minerals in the bottom and deep under the bottom of Black Sea, and more are expected to be found. In the first place we should mention the traditional inert building materials - gravel and sand. The big accumulations of similar inert materials are situated in the north-west shelf of the sea, mainly in the region of Odes. Exploitation works are carried out in several places and over 5 million m^3 of sand with low content of carbonate materials are dug per year (at a depth of 20 m). It is used in building, silicate and metallurgical industry.

The Bulgarian aerial shelf is narrower and considerably less in space in the north-west shelf of Black Sea. Here the content of gravel and sand in bottom depositions quickly decreases and the slimy depositions appear at a small depth. The reason is the absence of big rivers that could carry similar material.

The biggest accumulation of sand in our basin is the bed "Coketrice", established and mapped still in 1867 in the south-west of Cape Emine by an English hydrographic ship (the shoal bears his name). Today the bed represents a crescent-shaped body whose highest part is 16 m under the water surface. At the time of its first mapping the bed has raised to minus 9 m;

obviously, the sandy body has been gradually washed away in the course of time.

Bed reserves of sand, many-grainy and many-fragmentary clam valves, are more than 100 million m^3. The sandy body exploitation, situated only several kilometers in the south of Cape Emine, is impossible without performing of suitable profile and complex investigations. The idea of artificial piled up beaches in the zone of Sveti Vlas - Elenite seems attractive, but its realization may be performed only after long and labour-consuming expert ecological research.

In some regions of Bulgarian seacoast the sand alluvia are enriched with specific heavy minerals, containing iron, titanium, zirconium, uranium, thorium, etc. (magnetite, monazite, rutile, ilmenite, zircon, etc.). There are similar beds (deposits) in many local parts. Rather considerable deposits of ilmenite rutile-zirconium are connected with the outfalls of big river arteries from north-west and east parts of sea. The richest deposit sections contain on the average 15 kg ilmenite in 1 cub. m of mass and several kilograms of zircon.

Magnetite deposits meet mainly in the foots of the Balkan, Caucasus and Pontian mountains. There are titanium-magnetite sands in the Turkish seacoast in the east of Bosporus.

There is a well localized deposition bed in the west periphery of Black Sea basin, in the interior of Burgos Bay. Here, two ore-bearing sections - Aheloy and Sarafovo have been investigated. The content of ore mineral magnetite is 2-3 % average in sand.

It is known that in the basin of Black Sea there are undiscovered beds of excavated fuels. Beds of oil and gas are found in the north-west part and opposite to the Rumanian seacoast at a depth approachable for sounding.

Rumanian geologists and geophysicists have been occupied in the search of oil and gas in the basin of Black Sea from 30 years. During the last decades Rumania invested a lot of funds for search and study of oil and gas beds in the region of Black Sea shelf. Rumania became one of the few countries in the world with own production of platforms for sounding in sea conditions. They are installed in Galaz and draw out by tugboat along the Danube to Constanza. 7 platforms for sounding were prepared to the end of 1990. These platforms have working areas of almost 3 sq.m. The platform is "carried" by legs columns 122 m in height which step over the sea bottom.

The whole construction weighs almost 10000 tons and allows work at a depth to 90 m, as the sounding chisel may penetrate to 8000 m under the sea bottom. The Organization "Petromar", based in Constanza, has found two oil beds.

A part of the "shelf" oil bed has been known in Bulgaria since 1951, situated in the region of Tyulenovo village and Kamen bryag. To study its sea part, a metal scaffold bridge was erected. The oil bed lies at a small depth (360 m) and is an excellent raw material for lubricants of especially high quality. The exploitation of the bed still continues but few soundings work with minimum capacity.

To our regret, this first find has not been followed by others. The Bulgarian geophysicists and geologists have mapped the regions of shelf that are of interest and may prove to be productive. Around 1980 a sounding of some perspective parts has begun with the Russian platform "Sivash". The obtained results have not been published.

Recently the Bulgarian government undertook steps to estimate the oil and gas perspective of the shelf by offers for concession plots. Well-known and powerful firms and companies as Statoil, Enterprise, Maxus, Texaco, British Gas, Ajip and others compete over the right to carry out investigations in 6 sections of our seacoast with own technique and means and with participation of Bulgarian specialists.

Except the fluid (gaseous and liquid) excavated fuels, there are solid phase products in nature as coal, for example. Lately, an other group of natural compounds troubles the imagination of scientists and specialists in the field of energetic These are gas hydrates. Gas hydrates are non-stohiometric (with variable content) and cell structure system of individual gas components and 6-17 molecules of water. By appearance and chemical properties the gas hydrates resemble ordinary ice. All gases having temperature of boiling under $+60°$ C and not coming into chemical reaction with water may compose hydrates. Methane hydrates were found in layers not deep in earth surface in the north regions of Russia in 1970. In 1974 methane hydrates of fine crystal form were found in young Black Sea depositions, and in the middle of 1988 geologists of the Russian Institute of Oceanology, Moskow, managed to take massive samples of methane hydrates when shipping on board of "Evpatoria", lying in deep-water depositions from the north part of Black Sea. According to some investigations, at a large area in the western part of sea a horizon is traced reflecting the sound waves, situated at 80-100 m under the boundary "water - bottom". If connected with the spreading of methane hydrate formations, then good perspectives for gas fuel may reveal before Black Sea countries. Gas hydrate formations decompose slightly at temperature increasing and decrease the pressure of the initial components, as one volume of solid phase may release to 200 volumes of gas. Under the gas hydrate horizon, which is hard to penetrate for liquid and gas fluids, natural gas - methane may be piled up. Whether gas hydrate formations will be the fuel of XXI century as some scientists maintain, near future will show.

In conclusion, we shall note that the search and extraction of oil, gas and others excavated from the sea bottom is still a new activity for mankind and have not achieved the necessary measures, reliability and ecological safety for environmental protection. That is why, all these activities in Black Sea should be carefully allowed and controlled. In all cases before starting works, the self-cleaning potential of the region should be evaluated, corresponding norms should be determined for permissible pollution and a system for control and measures should be built against ecological consequences in accidents as well as in normal exploitation of the equipment's in sea.

We should always remember that Black Sea is a closed sea!

3

WHAT IS GOING ON TODAY IN THE BLACK SEA?

3.1. HYDROPHYSICS

Black Sea has taken its today's pattern and outline about eight-ten thousand years ago. That time coincides with the last warming in the Mediterranean and withdrawal of the glacial cover.

Its area is 423000 km^2, volume - 538124 km^3 and mean depth - 1272 m.

Water Balance

Rivers transfer 338 km^3/year fresh water in Black Sea, 237.7 km^3/year water fall as rain and snow, 176 km^3/year salty Mediterranean water run through the Bosporus by the bottom stream and 49.8 km^3/year water come from the Sea of Azov through the Kerchen strait. Annually 395.6 km^3/year of water evaporate, 371 km^3/year run into Marmora Sea through Bosporus by the surface stream, 33.4 km^3/year flow out in the Sea of Azov through the Kerchen strait. Therefore, the positive water balance of Black Sea is about 1.5 km^3/year.

For comparison: Burgos Bay, limited by the line Pomorie- Sozopol, contains about 2 km^3 water.

Balance of Salt

Annually 6326, 593 and 93 million tons of salts come into Black Sea from Marmora Sea, Sea of Azov and the rivers, correspondingly. At the same time the sea loses 6484 tons of salts annually in Marmora Sea and 568 million tons of salts in Sea of Azov, i.e., the sea becomes less salty with about 40 million tons of salts every year.

Heat Balance

Annually the sea receives 1998×10^{12} million Joules and due to its own emanation, contact heat exchange with atmosphere and evaporation, it loses 1972×10^{12} million Joules. In short, we observe warming of sea with 26×10^{12} million Joules per year.

For comparison, thousand Joules heat increases the temperature of one liter of water with 0.24 degrees.

Climate

The climate of Black Sea is determined by the geographic position (the annual quantity of heat) and by the prevailing north-west transfer of air masses.

Continental air prevails over Black Sea in the course of 185 days, tropical - 87, sea polar air - 50 and arctic - 43 average per year.

This determines the quantity of sunny days in the different parts of sea and rainfalls: more sunny days and less rainfalls in the west part and less sun and much rain in the east. Rainfalls in the west part are between 300-700 mm. and in the east 2200- 2700 mm., i.e. about six times more.

The mean annual temperature in C° is as follows: Burgos - 12.3°, Odes - 10.3°, Yalta - 13.1°, Novorosiisk - 12.9° and Batumi - 14.5°.

Due to the mean temperature increase in the north hemisphere and the intensified atmosphere circulation in recent years, a decrease is observed of the number of days with stormy winds and heavy sea. This decreases the wave mixing of the active sea water layer and worsens the utilization of harmful substances in it.

Hydrology Level

The level of Black Sea has annual variation of about 30 cm. In spring the level increases and in summer and autumn decreases. In shelf regions the level also changes in the limits of 20 cm because of the wind. The change of level may reach to meters where it is shallower, as in the Odes Bay.

According to data for the period 1875-1985 an increase of the water level is observed, as in the last years the mean increase is 1.5-2 mm/per year. This conforms to the global tendency of increasing the World Ocean level due to warming of planet.

Saltiness

Saltiness is measured in parts pro million (per mil) - milligrams salts per liter water.

According to its saltiness, Black Sea water divides into three layers. Saltiness in the upper layer changes under the action of river influx, evaporation, rainfalls and water from Sea of Azov - from 11 to 18 per mil. The changes of saltiness in the mean layer have seasonal character (from 15 to 22 per mil). The lower layer has formed because of the bottom stream through Bosporus with heavy salty Mediterranean sea water and changes by mixing with the mean layer (from 22 to 26 per mil).

The boundary between layers with different saltiness is called chaloclin. Thus, there are two chaloclins in Black Sea, as the depth of the upper one is determined by the seasonal temperature and for this reason is called seasonal or temperature chaloclin. The depth of the lower layer is determined by the ratio between the Mediterranean Sea salty and heavier water and the fresh and lighter water from rivers and rainfalls. Its depth in the central parts of sea is between 50 and 80 m and in coastal regions is between 75 and 165 m.

The presence of constant chaloclin, slow mixing of waters in depth, absence of oxygen in depth and appearance of hydrogen sulphide are unique hydrodynamic and hydrobiologic peculiarities of Black Sea.

Temperature of Water

Water temperature in Black Sea is also divided in three layers. Uppermost changing layer (at 200 m depth from the surface), where daily and seasonable, vertical and horizontal changes of temperature are observed. The intermediate layer is at 200 to 500 m depth, where the changes connected with the seasonable cycle of heat are thousand times weaker than these of the upper layer, but still vertical and horizontal seasonable changes are caught in the temperature. The deep water layer (from 500 m to bottom) is with almost constant temperature slightly increasing toward bottom. Sometimes, temperature inversions are established connected with the spreading of Bosporus water.

In coastal regions, at strong and longer wind from coast to sea, we observe upwelling - rise of colder waters from bottom to coast.

In summer the surface layer heats to 10 - 30 m depth. The boundary between warm and cold water is called thermoclin. Thermoclin is more stable than chaloclin in summer because of the bigger difference in water density. Thermoclin decomposes in winter and this intensifies mixing of waters in depth. The horizontal distribution of temperature on the surface has seasonable constancy of the temperature minimum in the north-west part of sea and maximum - in the south-east part.

The cold middle layer is a specific peculiarity of Black Sea. It is fed by water cooled in winter in the upper layers of circular streams in the central parts of sea, which flows down the cupola formed by rotation.

The analysis of centuries-old recordings of temperature shows cycles of warming and cooling with periods of 2 to 5 years, 20 and 60 years. The coincidence of cold or warm phase of these cycles leads to anomalous cold or warm years.

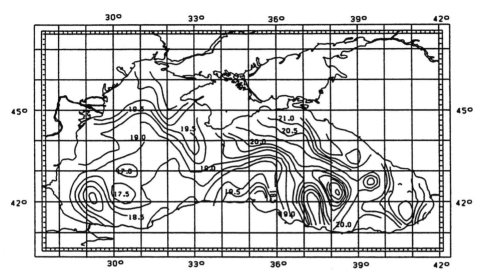

Figure 1: Satellite Monitoring of Surface Temperature

Density

Sea water density is proportional to saltiness and inversely proportional to temperature. We remind that in practice water is unbending at increasing the pressure.

More salty and colder water is heavier and, therefore, goes down. That is why pinoclin, i.e. the boundary between warmer and less salty surface water and salty Mediterranean sea water, is a barrier to mixing of waters in depth.

Streams

Sea ecosystems may function because of heat transfer, dissolved and suspended components. That is why the experimental and theoretical knowledge of horizontal and vertical streams is of prime importance.

Just at the end of the 70's, the scheme of streams in Black Sea has been studied in detail.

The general scheme of surface streams represents closed, mainly cyclonic (i.e. counter clockwise) rotations. The basic Black Sea stream (BBSS) encloses the entire sea along the end of shelf. Two arms stray from it: one from Cape Cham to Cape Pizunda and an other that strays at Cape Saric, enters the north-west part of sea, joins the Danube waters and interflows the BBSS at Cape Kaliakra. The mean velocity of BBSS is 0.3- 0.5 m/sec, as the jet reaches to 0.4-0.6 m/sec in the center. The maximum velocity reaches to 1.0 m/sec at strong favourable winds.

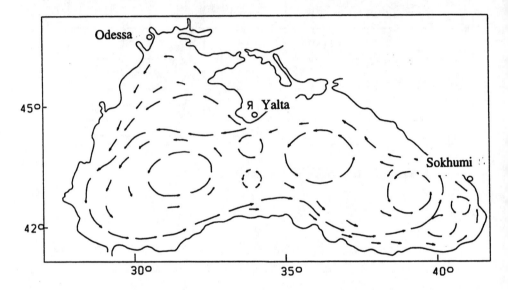

Figure 2: Average Annual Schemata of Quasi-stationary Surface Currents in the Black Sea.

It is easy to calculate that the stream passes 21.6 km average per 24h, i.e. 11.6 miles. This shows, indeed, that a pure Burgos Bay cannot be pure without a pure Odessa Bay.

In the deep water part of sea, surrounded by BBSS, cyclonic whirlwinds are observed as one of them, in the east part, is anticyclone. There may exist several whirlwinds in the west part of sea depending on the wind situation. Anticyclone whirlwinds are formed around the capes jut out deep into sea or around the great canyons of shelf in the west part. BBSS may change its direction at strong contrary winds as its maximum velocity reaches 0.25 m/sec.

Streams in the coastal part are parallel to coast and change their direction irregularly in a period of about 6 days, i.e. they have bimodal character. We underline this fact because it is important when determining the volume, content and regime of outfall of everyday wastes and industrial waters in proximity of coast.

Recently, chains of anticyclone whirlwinds were discovered and traced by observations from space between the coast and BBSS of lifetime 5 - 7 days and period 16 - 18 days. In spring and summer, when the stability of BBSS decreases and the formation of meanders increases, the number of coastal anticyclone whirlwinds may reach even four per month. In this manner, the existence of a convergence zone is established in the coastal part of sea. The existence of this zone is the main reason for a restraint of pollution in coastal waters, since it hinders the ventilation of the bays in a great period of the year. This zone is a natural hydrophysical basis of the biological boundary between the ecosystems of shelf and in open sea.

Specifying the picture in the zone of convergence will allow to estimate the ecological responsibility of local pollutants, the contribution of big rivers and the exchange atmosphere-sea.

The vertical component of streams velocity in the middle of coastal anticyclone whirlpools is pointed downward and coincides with waters in the periphery of cyclone rotations. This increases the intensity of going down of polluted coastal waters along the continental slope. As a result, the raising of bottom waters increases in the centers of cyclone rotations.

Thus the previous estimation of deep-water changing in the stream at 2-3 thousand years reduces to 100- 200 years.

This shows that deep-water outfalls and burial of polluting substances in the bottom of sea may turn into an ecological time-bomb.

The analysis of surface water going down, carried out on data based radioactive distribution after the Chernobyl accident, shows a tendency to shorten this term to 50-80 years. If this proves true, deep-water outfalls of pollutants dissolving with difficulty must cease immediately, otherwise a total ecological catastrophe will come before long.

The experimental determination of the vertical and horizontal transfer coefficients allows to solve different tasks for spreading of pollutants containing different components.

3.2. HYDROCHEMISTRY AND SPREADING OF POLLUTING SUBSTANCES

The spreading of physical admixtures, chemical elements and their inorganic and organic compounds in sea along the horizontal and vertical line is of interest with respect to their influence on the ecosystems.

The cooled surface water, containing dissolved oxygen, goes down to bottom in oceans, due to their high (about 32-35 ppm) and almost equal vertical saltiness. For the small concentration of the organic substance, the velocity of oxygen consumption is no less than the velocity of its entry. For this reason there is oxygen to its very bottom.

The saltiness of surface waters in Black Sea is 16-18 ppm. It increases to 20.8 promil in layers from 50 to 150 m. This is the layer with a jump in the density- pinoclin. The saltiness is about 22 promil from this layer to the bottom.

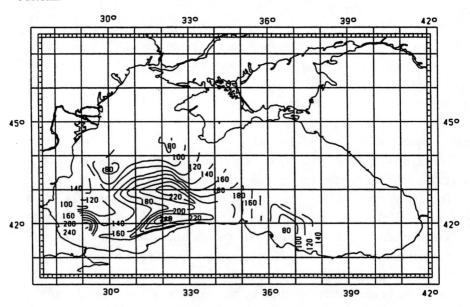

Figure 3: Spreading of Organic Compounds in the Western Part of the Black Sea (mkv/l) Nov.-Dec. 1989

Dissolved Oxygen

Oxygen enters into water from air and is produced by the phytoplankton. Water saturated with oxygen may be observed during blossoming.

Oxygen concentration in open sea in the layer to 0.5 m is from 6.13 to 7.85 ml/l, 90-112% saturation in spring; from 5.18 to 6.53 ml/l, 100-120 % saturation in summer; from 5. to 7.00 ml/l, 100-105% saturation in autumn and 6.96 to 9.75 ml/l, 94-106% saturation in winter. These average data are for 1963-1985 and are higher than those for the period 1939- 1962. This increase is explained with the accelerated blossoming of plankton due to pollution.

The average oxygen concentrations at horizon 50 m are from 2.54 to 8.48 ml/l (36- 110%) in spring, from 7 to 7.35 ml/l (51-102%) in summer, 3.07 to 7.15 ml/l (43-134%) in autumn and from 3.16 to 7.99 ml/l (51-101%) in winter depending on region. The concentration and unsteadiness of this horizon are higher than those for the period 1939-1962.

At a depth more than 60 m, the oxygen concentration sharply decreases and is about 0.5 ml/l at a depth to 100-110 m. Hydrogen sulphide appears at this depth. In the layer of simultaneous existence of oxygen and hydrogen sulphide, the oxygen concentration varies from 1 ml/l to analytical zero. Depending on region and season, related to cyclone or anticyclone whirlwinds, oxygen disappears at depths from 145-180 m to 150-280 m.

The analysis of oxygen spreading in years shows an unstable state tending to a concentration increase in surface layers. This is the effect of intensive anthropogenic impact on sea, warming of climate and increase of bioproductivity.

To obtain a more precise picture with models performing in real time, a completely new approach is necessary for organization and financing the investigations. Only by means of such precise picture it would be possible to forecast the behavior of ecosystems with reliable accuracy and in time.

Hydrogen Sulphide

Anaerobic conditions had appeared in Black Sea about 8-10 thousand years ago, when salty and heavier Mediterranean Sea waters had begun to penetrate through Bosporus and that had led to a hampered disarrangement of the water mass in depth. According to geological data and theoretical models, the level of hydrogen sulphide had increased toward the surface for about 5000 years and reached its stable present position about two thousand years BC. There are data for small variations of the level connected probably with the sun activity in the last 300 years.

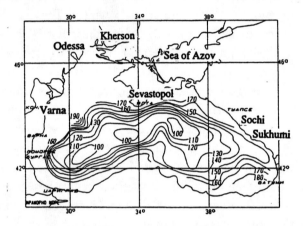

Level of Hydrogen Sulphide 1920-1930

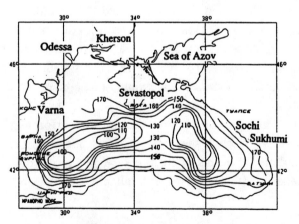

Level of Hydrogen Sulphide 1980-1985

Figure 4: Level of Hydrogen Sulphide in the Black Sea in the 20th Century

Lately, some scientists, journalists and statesmen talk about an increase of the hydrogen sulphide level, even about flowing through Bosporus. Their analyses ignored the strong unsteadiness with time and the uncontrolled increase of accuracy of the batometric sample in the last fifty years. The recording of these factors does not allow one to draw a reasonable conclusion about the hydrogen sulphide level raising.

The precise analysis of existing data shows practically an identical picture of hydrogen sulphide level in Black Sea for the periods 1984-1986 and 1924-1927. Simultaneously, a tendency is observed toward a concentration increase at different depths analogous to the increase of oxygen concentration in the upper layer. The layer thickness of oxygen and hydrogen sulphide co-existence increases from several ten meters some thirty years ago to 50-180 m in present time. This is explained with the increasing anthropogenic impact that leads to an increase of the organic substance in the anaerobic layer of sea, climatic (warming of climate) and geological causes, increase of the hydrogen sulphide production in depth. Simultaneously, the increased content of oxygen in the surface layer is enough to oxidize the hydrogen sulphide. In open sea, the oxygen concentration is sufficient to oxidize the hydrogen sulphide, coming out of sea depths before reaching the surface.

In all cases the hydrogen sulphide concentration, produced in the water column, is so low that its self-ignition is impossible in air over the sea surface.

The observations of fires in sea during the Crimea earthquake is explained only with the release of natural gas and its self-ignition because of the great gradients of earth electrostatic field which is an often met phenomenon.

For example, the earthquake in Vrancha in May 1990 released hydrogen sulphide from the old antiseptic pits of Pomorie and the smell of hydrogen sulphide was everywhere in town. Our third seminar was held there in that time.

The hydrogen sulphide concentration in the layer of co-existence with oxygen varies from traces to 0.5 ml/l. After that the concentration increases and reaches to about 4 ml/l at a depth of 250-500 m. Further on the concentration changes toward the bottom from 6 ml/l to 7.6 ml/l., as the variations are connected with regions of breaks and processes of distribution of Mediterranean Sea water.

Still there is not one simple concept about the mechanism of hydrogen sulphide yield. Most possibly hydrogen sulphide is obtained as a result of biological anaerobic disintegration of the organic substance. A considerably smaller part may be obtained following the abiogenic sulphate-reduction. Geological origin is also possible in small quantities at several places.

Lately the influence of sun activity and earth crust motions on the velocity of generating hydrogen sulphide are studied in depositions at the sea bottom.

It becomes clear that the depths of Black Sea are one gigantic factory for yield and conversion of hydrogen sulphide working ceaselessly already about 4000 years.

In our opinion, the present technological possibilities and the price of sulphur in the international market do not allow ecological and efficient production. Proposals for cleaning the sea from hydrogen sulphide by

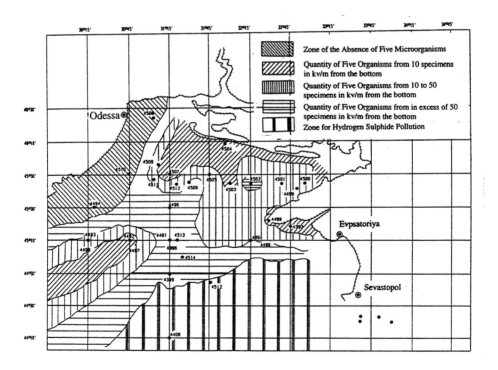

Figure 5: Pollution of the North-western part of the Black Sea as of 1988

production of sulphur are utopian plans of people who cannot imagine (maybe this is unprofitable for them?) how large four hundred thousand km^3 water is.

In fact, the serious dangers are shelf mud and water toxicity as a result of hydrogen sulphide pollution of mud in the seacoast. This process will be explained in the biological chapter. Here only its velocity will be outlined. A sharp decrease of the phillophorn field of Zernov was observed till 1985 in the northwest part of sea.

For the first time in 1986, a continuous blossoming of plankton has been observed in front of the west seacoast.

The first islands of slime on the bottom were observed in 1987 in Odes Bay. In the autumn of 1989, the whole bottom from the Danube delta to the mouth of Dnieper at a depth of 25-30 m was covered with a layer of slime from 10 to 80 cm smelling of hydrogen sulphide, while the layer of water at a depth of 2 m over bottom was toxic and oxygen free.

Since 1989, the beaches of Odes were officially forbidden for swimming and consumption of scad caught in the region of Odes Bay leads to serious intestinal disorder. With a delay of several years, the anthropogenic impact shows its action in Burgos and Varna Bays.

Chlorinated Hydrocarbons (gammexane)

About 860 analysis has been done for the period 1975-1985. The results are given in seasons.

Winter: Surface waters contain more than 1 nanogram per liter (ng/l) everywhere. The maximal values are from 22 to 28 ng/l in coastal regions with an increased influx of river and rainwaters. At a depth of 10 m, there is a decrease of the concentration, as the minimal values are about 2 ng/l, the maximal are 9.6- 9.9 ng/l in the same coastal regions. The decrease reaches to 50 m depth, where the maximal concentration reaches to 9 ng/l.

Spring: Almost the same concentrations, a slight increase is observed after rains. Data show absence of strong dispersion in depth.

Summer: Increase of surface concentration to 8.2 ng/l. For horizon 10 m the maximal concentration is 13- 23 ng/l. At 50 m in the coastal zone it is 10- 15 ng/l and no measuring in open sea.

Detergents (Synthetic Surface Active Substances (SSAS))

For the period 1973-1985 the average concentrations of SSAS change up to 38 micrograms per liter (mkg/l) in spring; to 83 mkg/l in summer; to 166 mkg/l in autumn and to 9 mkg/l in winter. There are maximal concentrations to 400 mkg/l at 10 m in coastal regions. The average concentration reaches 760 mkg/l at 50 m. The increase of concentration from spring to autumn coincides with the increased pollution during the tourist season. The increase with depth is a fact that reminds about the necessity of new technologies.

Oil Hydrocarbons (OHC)

The sea area, polluted superficially by OHC, has increased 5 to 15 times in 1988 compared to 1987. The average concentration of OHC in layers at 0-5 m varied from 0.05 to 0.41mg/l in 1988. The layer at 0 - 100 m has contained 0.1 mg/l for the period from 1978 - 1988.

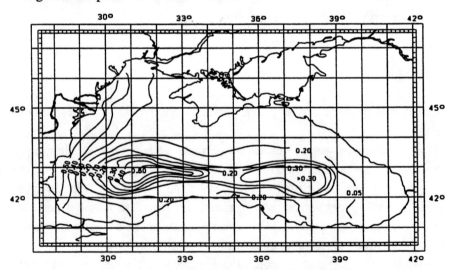

Figure 6: Distribution of Petroleum Production on the Microlayer (mg/l)

The permissible limit concentration (PLC) for OHC is 0.05 mg/l. If the pollution with OHC decreases with 10% per year, PLC will be reached in four years in the superficial layer.

The concentrations of OHC exceed tens of times the PLC in bays. A typical example are the harbour bays. In America, the estimation of damages caused on sea by spilled oil products is 250 dollars per liter.

Silicon Phosphates

After 1970, when building of dams on rivers emptied in Black Sea was finished, the content of silicon and phosphates in sea water decreased from 2 for silicon and to 10 times for phosphorus in the photic layer. Their concentrations stabilize for silicon from 70 to 520 mkg/l in the period 1980 - 1988, and for phosphorus from 6 to 18 mkg/l.

Ammonium Nitrogen

Up to 80-100 m, there are two minima of concentrations at depths of 10 - 30 m (zone of photosynthesis) and 70-100 m, where the concentration falls to 0 - 0.2 microgram-atom per liter(mkg-atm/l), almost no oxygen and traces of hydrogen sulphide appear. Then the concentration of ammonium increases and exceeds 1.5 mkg-atm/ l at 200 m. At 500 m it is already 100 mkg-atm/l. This vertical structure preserves along the whole main Black Sea stream. In the centers of cyclone whirlpools, where upward streams exist, the concentration reaches to 10 mkg-atm/l at a depth of 80 m. The increase of ammonium nitrogen concentration is an indication for intense anthropogenic impact on the coastal regions. In the bottom waters of the northwest part, the concentration varies from 37 to 550 mkg-atm/l.

Urine (uric acid)

There are two maxima of concentration in open sea: at depth of 5 - 20 m and 60 - 90 m - from 1 to 2 mkg-atm/l. The concentration of urine in the anaerobic zone is almost zero. Around the outfalls of the Danube, Dnieper and Dniester, the concentration reaches to 5.0- 8.0 mkg-atm/l. In muddy waters the concentration increases at least ten times.

Superficial Microlayer (SML)

This is the surface layer thickness of not more than 1 millimeter. Although thin, the qualities of this layer are significant because the evaporation and penetration of atmosphere into water passes through it. Its properties are strongly influenced by the presence of oil hydrocarbons.

Its saltiness changes more strongly than that of the one-meter layer and may serve as a standard for the river influence. The quantity of bionic substances is 100 - 1000 times more than that in the one-meter layer. The quantity of nitrates and nitrites in SML is highest in coastal regions, as their ratio is 4:1. The oxidation is determined by the different components of the dissolved organic substance (carbon, oxygen, hydrogen, phosphorus, sulphur, potassium, nitrogen, calcium, etc.) and is highest near the big towns. The concentration of mercury in SML is about 1.0 mkg/l.

The anthropogenic impact after the 70's strongly changed the properties of SML. This embarrasses a lot the estimations and, therefore, the prognoses for self-cleaning capacity of sea.

3.3. RADIOACTIVITY BEFORE AND AFTER THE CHERNOBYL ACCIDENT

It may sound cynical, but should be said, that the Chernobyl accident helped scientists to get organizational and material possibilities to investigate the background radioactivity and its influence. Furthermore, the ecological knowledge of people considerably increased and conditions were created for an essential control over the distribution of radioactive materials in the Black Sea water collecting zone.

This does not mean that scientists are pleased with the catastrophe. On the contrary, as a rule they have warned and tried to take measures years before the accident. Unfortunately, people begin to listen too late.

Radioactive elements exist on Earth since the time of its formation as planet (isotopes of uranium U-238, U-235, thorium Th-232, radium Ra-226, potassium Ê-40 and their by-products), as a part of them come as a result of cosmic rays interaction with atmosphere and earth materials (carbon C-14, beryllium Be-7, etc.). Radioactivity had been much higher and the set of radioactive nuclides richer during the formation of planets. For these four billion and a half years of existence the comparatively short-lived nuclides had disintegrated. In the last two billion years the radioactive background had slightly varied because of changes in the sun activity and intensity of cosmic rays along different parts of the solar system orbit in our Galaxy.

That was the picture till 1945. After this date the technogenic radioactivity of the planet has been increasing continuously. The number of radioactive nuclides is increasing too. The biological influence of this increase may be already measured and, of course, not in favor for man.

Initial source of radioactivity were the barbarian tests of nuclear and thermonuclear weapons in atmosphere, oceans and underground. Only France and China still continue with such explosions.

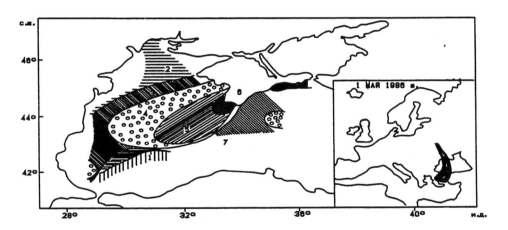

Figure 7: Chernobyl Catastrophe Radioactive Cloud on May 1, 1986

Distribution of Cs^{137} on the Surface of the Water (1986)
1 – min. to 100 Bq/cub.m.
2 – from 100 to 200 Bq/cub.m.
3 – from 200 to 300 Bq/cub.m.
4 – from 300 to 400 Bq/cub.m.
5 – from 400 to 500 Bq/cub.m.
6 – from 500 to 600 Bq/cub.m.
7 – above 600 to 100 Bq/cub.m.

After the Agreement of 1963, the entrance of radioactive materials in the environment cut down sharply but unfortunately did not stop. First, not all countries in the world joined the agreement. Second, the development of atomic energy, the use of radioisotopes in industry, medicine and science, the maintenance of an enormous arsenal of nuclear and thermonuclear weapons create conditions for an incessant increase of background radioactivity in the environment. The radioactivity in Black Sea is an illustration.

According to measurements of strontium Sr-90 for the period 1960 - 1973, radioactive materials in quantities of 615 Ci have fallen in Black Sea in 1960 (1 Ci (curies) is equal to 37 billion decays in a second), 4403 Ci in 1963, with reduction to 520 Ci in 1973. For the period 1953 - 1959 a gradual decrease has been noticed of the quantity of radioactive materials in atmosphere, while in February 1960 a hundred times increase has been observed because of the French nuclear explosion in Sahara on 13 February

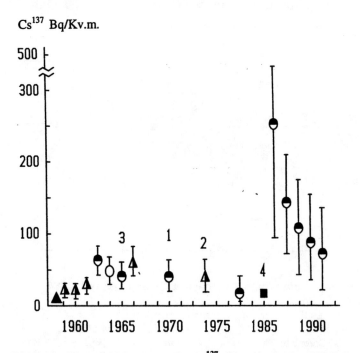

Cs^{137} Bq/Kv.m.

Figure 8: Historical Data on the Cs^{137} Content in the Black Sea

1960. The maximal contamination of Black Sea with radionuclides was in January 1962 as a result of renewal the atomic and thermonuclear bomb tests with new technology - 660 mCi/sq.km. The total activity fallen in Black Sea from atmosphere for the period 1960- 1974 is estimated to 2.5 million Ci.

The contribution of the Danube for the radioactive contamination of Black Sea in the period 1960-1968 is estimated to 2000 Ci. As a consequence of building atomic industrial objects along the Danube and the absence of efficient international control, this increase began to enlarge drastically in the last years.

The spreading of caesium Cs-137 in the upper hundred meter layer of sea-water is uniform - about 17 Becquerels per cub. meter (Bq/cub. m) in 1977. The total quantity was about 42000 Ci.

As a result of the Chernobyl accident, radionuclides have fallen into the environment with total activity above 50 million Ci (more than 30 radioisotopes) in the form of gases, aerosol particles of micro-particles from the reactor fuel. The content of radioactive contamination changed quickly due to the different lifetime of its components. For example, the dangerous for man iodine J-131 (period of half-decay 8 days) was absent from the release products a year later. Even after hundred years, ten percent of Cs 137 will remain in the environment, because its period of half-decay is a little more than 30 years. The long-lived isotopes of caesium, strontium, plutonium, americium will remain for years to come.

Immediately after the accident, a radioactive cloud turned to Scandinavia, but on 30 of April the wind changed and on 1 of May the radioactivity began falling into Black Sea. Practically, the whole East Europe and a considerable part of Western Europe were contaminated more or less. Since Black Sea collects waters from the entire region, the technogenic radioactive substances will collect there.

The spreading of radioactivity in surface waters corresponds to our knowledge about streams and atmosphere dynamics. The eastern part of sea proved to be most strongly contaminated. The route of the cloud was traced - through West Crimea toward the Bulgarian south seacoast. Estimating data for caesium isotopes in surface waters in the west of Crimea, there were 1090 Bq/cub.m (before the accident 15-17 Bq/cub.m) and about 450 Bq/cub.m opposite to the Bulgarian coast.

In the end of 1986, the spreading of caesium isotopes was much more uniform and covered the entire basin of sea. In the south-west part of sea (40-60 miles east of Tsarevo), a clearly pronounced maximum was observed by Cs 137 - more than 400 Bq/cub.m. The explanations for this maximum are two for now. The first one is that those were waters from the Crimean coast in the summer of 1986. The second one is that there was some local source in this region. Our additional investigations in May 1990 showed that if there was any, it was not in the bottom.

The total quantity of Cs 137 in the upper 50 meter layer was 110000 Ci in December 1986 and decreased more than twice in 1987. This is a good estimation for the capacity of Black Sea to "self-clean" by moving of radionuclides into biological objects or by deposition in the bottom.

The other radionuclides (isotopes of plutonium, americium, strontium...) give total of 110 to 700 Bq/cub.m. The isotope Sr-90 is sorbed by the substance suspended in water much more than caesium and is especially dangerous, because is similar to calcium by properties.

The investigation of the background radioactivity after the Chernobyl accident gives grounds for a new and rather serious alarm. The measurements and the preliminary modelling of caesium isotopes distribution after the accident showed that several times more radioactive products have fallen in Black Sea in unknown ways than as a consequence of the Chernobyl accident for the last ten years. In short, there are reasons to consider that the barbarian secret throwings of radioactive wastes in Black Sea go on. Our opinion is that only an Association of Black Sea Towns and Regions "Clean and Peaceful Black Sea" will be able to create for a short time an active system for control of the sea from such malign contaminations.

After the Chernobyl accident, scad contained from "traces" to 50 Bq/kg, Sprattus - to 50 - 62 Bq/kg, mussel meat from 10 to 40 Bq/kg, green weeds - to 50-70 Bq/kg and brown weeds - to 230 Bq/kg along the Bulgarian seacoast. These values mean an increase of one and a half to three times.

Interesting are the data for Black Sea mussel in the region of Pomorie, whose meat did not increase its radioactivity in practice, unlike the shell whose radioactivity reached to 100- 130 Bq/kg.

As far as we know, nobody has investigated systematically the radioactivity distribution in different organs of fishes of trade importance and of dolphins as well. Such an investigation should give a possibility to reduce the eventual risk of fish consumption and benefit to a better understanding of the causes for increased death among dolphins.

In the spring of 1990 the radioactivity of the 50-meters water layer by caesium Cs-137 was from 15 to 75 Bq/cub. m. and by strontium Sr-90 the mean activity was about 13.7 Bq/cub.m. Maximal radioactivity, equal to 105 Bq/cub.m, was measured near Stavrova bank in Burgos Bay.

Chernobyl added to the radioactivity of shelf in the west part of sea. Cobalt Co-60 is observed in some places. The mean content of uranium U-238 in surface waters of the Bulgarian seacoast is 2.0 mg/ton, as in some points of Burgos Bay reaches to 2.8 mg/ton.

3.4. ANTROPOGENIC EUTROPHICATION OF BLACK AND MEDITERRANEAN SEAS

The biggest role in the anthropogenic influence on seas and oceans plays the increasing influx of bionic substances. This process generates numerous interrelated phenomena in natural basins joined by the term "syndromes of eutrophication". Including "blossoming" of water or, according to Yu. Odut, a "malign" increase of biological productivity followed by oxygen deficiency in bottom water layers (hypoxia) and mass death of bottom organisms, release of hydrogen sulphide in the process of decay of albuminous substances, decrease of water transparency, sliming, etc.

Eutrophication of basins, and also of seas and oceans, got an especially fierce development during the last 2 - 3 decades as a result of intensification of agriculture, industry and other man's activities. The extent of provoked atrophy of every separate basin depends on the specific physic-geographic, hydrological and hydrobiological conditions.

Black Sea, as an object of anthropogenic eutrophication, has been studied since the end of the 60's. Gradually, the concentration of nitrogen and phosphorus compounds and the mass of phytoplankton, especially of pyridine weeds, increased in its north-west part where the three biggest rivers - the Danube, Dniester and Dnieper flow. These processes have been developing quickly and in the 70's the "syndromes of eutrophication" were clearly expressed: "red tides" and death of bottom organisms.

Numerous articles appeared in literature about the complex and even critical state of Black Sea ecosystem, especially in the north-west shelf where the situation compared to the beginning of the 60's changed basically. A systematic research has begun of the entire Mediterranean Sea basin with the efforts of many scientists.

The water surface of the Mediterranean seas, including Black and Azov seas, is about 2.96 million sq.km. The basin collecting the water of these seas in the territory of Europe, Asia and Africa exceeded 7 million sq.km. Therefore, the size of the specific catchment area (the ratio of the catchment area surface in land toward the surface of sea) is close to 2.4. This index deviates significantly from the mean value for the separate seas. Thus, for the Sea of Azov it is more than 19.0, and for Black Sea – 5.6, Aegean – 0.7, Adriatic – 2.1, Ionian – 0.3, Tirenian – 0.4. The specific catchment area reaches 2.6 for the rest parts of Mediterranean Sea.

The specific catchment area is not, of course, the absolute and unique criterion for estimating the dependence of sea on land. In present days, however, rivers and sea towns are exactly the main source of anthropogenic pollution of sea ecosystems.

To draw up a map on the pollution of Mediterranean seas on the basis of chlorophyll "a" concentration in the surface layer of pelagian, data have been used by distant measurements of cosmic devices and by traditional methods from ships and buoy measuring systems. The information received from satellites contains less parameters, but in practice is instantaneous. For example, a two-minute cosmic picture gives two million images by means of the device CZCS (Coastal Zone Color Scanned) from satellite "Nimbus- 7" of NASA, covering water surface equal to 2 million sq.km. More than 11 years would be necessary for the same work on ship expedition with velocity of 11. At present, the combination of data from distant and contact measurements is most informative.

The cosmic photo of the Mediterranean Sea was taken in May 1980. According to observations, the process has advanced essentially in the last 10 years in separate regions.

A great part of the Mediterranean Sea is oligotrophic at present with minimal quantity of phytoplankton. This is typical for the eastern half of sea. Lately, there are regions of high atrophy along the coasts of Catalonia, Provance, Cote d'Azure, the Italian Riviera. Concerning Adriatic Sea, the situation has been critical along its north and west coasts since the middle of the 80's. A main source of eutrophication is river Po, crossing a territory of intensive agriculture, cattle-breeding and industry. In Alboran Sea, mineral and organic fertilisers enter with surface stream waters from the near part of the Atlantic Ocean. On the opposite northeast coast of the Mediterranean Sea, in the region of Iskenderon Bay, a zone of high atrophy is formed by the rivers Jahan and Sahan. The vastest eutrophication of Mediterranean Sea water-basin is in the outfalls of rivers and neighbouring waters, where bionic substances of farming and industrial origin infuse. Wastewaters from big towns and villages along rivers downstream also supplement. A comparatively small number of aquatic basin of Mediterranean Sea eutrophicates not from rivers but from unclean wastewaters of towns, harbours and other coastal local sources.

If about 473 cub. kg of river water per year empty in Mediterranean Sea, with its area of 2.5 million sq.km and volume of 3.7 million cub.kg, then in Black Sea, with area of 422 thousand sq.km and volume of 547 thousand km^3, a river flow of about 350 cub.kg empties average, including from the Danube 203 cub.kg per year at a moderate vertical shifting of water masses and limited water exchange through the Bosporus strait.

Sea of Azov with area of 39 thousand sq.km and biggest depth of 13 m in high water years receives more than 40 cub.kg river waters. For this reason the Sea of Azov experienced the "syndromes of eutrophication", including the mass deaths of bottom organisms, and in previous decades - total eutrophication of coastal sea waters. However, deaths included only the central part of sea. Nevertheless, it remained one of the richest among the salty seas in the world according to its biological and fish productivity.

By their extent of anthropogenic eutrophication, the separate seas can be ordered in decreasing series, as follows: Sea of Azov, Black, Adriatic, Marmora, Lion Bay, Aegean, Balearic, Alboran, Ligurian, Tirenian, Ionian, Levantine, the central parts of the Algerian- Provance basin and the eastern half of Mediterranean sea.

The incessant area of water "blossoming" is periodical in summer along the entire basin of Sea of Azov (till 1988), about ¼ of Black Sea surface, the northern half of Marmora, about 1/5 of Adriatic and much less in the other seas.

The vastest hypertrophic aquatic basin in the Mediterranean is situated in the north-west part of Black Sea (along the seacoast of Ukraine, Rumania and Bulgaria in the zone of influence of inflow from the Danube, Dniester and Dnieper). Its area is to 100 thousand sq.km. Here the biggest values were measured in numbers and the biomass of plankton organisms reacting positively to eutrophication of sea water. So the number of phytoplankton (with prevailing of pyridine) to 800 million sample/l in biomass of 1000 g/cub.m, and infusorii Mezodinium rubrum correspondingly 4.6 million sample/l and 280 g/cub.m. As for Noctilucamiliaris, in the layer of neston (0-5 cm) it may reach to extreme values for the World Ocean, exceeding respectively 6800 million sample/cub.m and 500 kg/cub.m. Also, in the north-west part of Black Sea, the highest number has been noticed of the new settler from the Atlantic Ocean - ctenophore, defined firstly as Bolinopsis infundibulum and later as Mnemiopsis leidyi. The number of young specimens of this organism in coastal waters may exceed 500 sample/cub. m.

The result of "malign" productivity were deaths on a large scale of bottom organisms that were observed on vast areas in the north-west shelf of Black Sea in the beginning of the 70's. For example, they cover an area up to 10 thousand sq.km with a depth of 35-40 m only before the coasts of Ukraine and Crimea. As a result of hypoxia organisms die at every square kilometer of bottom from 100 to 200 tons, including 10 - 15 tons of fish - industrial and non-industrial, full-grown and young specimens. Deaths cease with autumn cooling and the increase of turbulent disarrangement of waters. The devastated regions of shelf began to be populated by larvae brought with streams from the spared regions.

Except losses in fish industry, the anthropogenic eutrophication of sea waters with its ecological consequences leads to large losses in other fields of economic and social spheres. This refers especially to the recreation activity of seacoast. The "blossoming" of water changes its habitual and acceptable for man organoleptic properties. Water becomes brown or grey, mucous in touch, with unpleasant smell. Aerosol particles enter into atmosphere from "blossoming" sea water, containing toxic substances and pathogenic micro-organisms.

The increased turbidity of water as a result of the mass reproduction of phytoplankton reduces the lighting of bottom and worsens or excludes the possibility for development of bottom weeds - macrophytes, which are a basic generator of oxygen in the coastal zone of sea and nutrition basis for many animal species.

The increase of concentrations of organic substance in water, the weakening of sun radiation, shortening the number of animals - filters and other water "health officers", as a consequence of deaths, create conditions for successful reproduction in water and bottom of pathogenic micro-organisms. This is already a direct menace for human health in contact with sea.

The anthropogenic eutrophication has enveloped many coastal regions, traditional for popular tourism, recreation and talasotherapy: the north-west resort region, the seacoast of Rumania and Bulgaria. In other places of the Mediterranean basin, complicated is the position of the north coasts of Marmora sea and the Prince Islands, some bays of Aegean Sea, the north and west coasts of Adriatic, the coasts of Marseilles in France to Valence in Spain and some others.

When discussing the wide spreading of eutrophication of sea waters, attention should be paid on the essential obstacle that water, containing a great quantity of plankton, has a temperature on the surface of 2-3 degrees C higher than water poor of plankton (under the same conditions). It is also known that the organisms of vegetable and animal plankton by moving their palpi accelerate the evaporation of water 2-3 times. In this way, the number of ecological consequences from eutrophication also added to the mechanism of climate formation processes.

Summarizing, it should be underlined that Black Sea and Sea of Azov are actually the most strongly eutrophicated aquatic basins from the entire Mediterranean sea basin. Serious ecological, economical and social problems appear in these regions that are not as much determined by the processes running in the seas alone, than the processes running along the entire catchment area. This is a territory of more than 2.3 million km^2 for Black Sea.

Local pollution of bays and coasts from local sources, even though strongly affecting the ecological situation (to an ecological catastrophe in some places, as in Odes Bay), still have not sufficient capacity to influence significantly on the ecosystem in open sea.

The accelerated death-rate among dolphins is a serious warning of unfavourable changes that already have begun in the entire Black Sea ecosystem.

3.5. HYDROBIOLOGY

Molismology of Black Sea

Molismology (of Greek origin *molismos*, means dirt, pollution) is a science about pollution's, their state and biological function.

Depending on their concentration and function on the ecological systems (in spite of the origin: natural or technogenic), pollution's divide into two types:
1. Pollution's with substances of concentrations above the natural background level, lower than the limit values that do not influence considerably on the ecosystem functioning;
2. Pollution's with substances of concentrations above the limit values that change considerably the ecosystem functioning and, on its side, this changes indirectly the state of pollutants.

Lately, the process of anthropogenic eutrophication of aquatic ecosystems reached to threatening and catastrophic dimensions (particularly for the shallow coastal waters of Black Sea).

Phosphorus content in waters is one of the limiting factors for eutrophication of the Black Sea ecosystems. The biochemical cycle of phosphorus in Black Sea has some peculiarities because of the hydrophysical and hydrochemical structure of waters.

The study of the horizontal and vertical distribution of phosphates in sea shows that the concentrations of phosphates (phosphorus as element) in depth is the following: upper oxygen containing layer - from 0 to 40 mkg/l, intermediate layer - from 10 to 140 mkg/l, hydrogen sulphide deep water layer - from 140 to 250 mkg/l.

The main sources of phosphates for the upper 100-meter layer are waters from river influx and phosphates from bottom waters. With river influx waters had entered: - 16.7 thousand tons/per year in 1952-1959 (from the Danube - 12.64 thousand tons/per year, from the Dnieper- 3.5 thousand tons/per year); - over 50 thousand tons/per year in 1981- 1985; with bottom waters had entered: 13-17.6 thousand tons/per year.

For the change of the hydrochemical indexes certify the many-years investigations of the bionic elements content in waters of the Bulgarian seacoast. The mean annual concentrations of phosphates in the north from Varna Bay were 39.4 mkg P/l for the period 1962-1975 and 77.5 mkg P/l for 1976-1980.

The mean annual concentrations of ammonium nitrogen were 56.0 mkg P/l for the period 1962-1975, and 71.0 mkg P/l for 1976-1980.

The increased anthropogenic chemical loading changes the biochemical content of hydrobionts and bottom depositions, leads to a change in species content of hydrobionts, to worsening the qualities of coastal regions waters.

The questions for efficient preservation of natural waters and ecologically approved measures for use stand more sharply.

The problem of sea pollution with polychlorinated aromatic and aliphatic compounds, including biphenyl, naphthalene, exists only since several decades. Polychlorine biphenyls (PCBP) in water are on the surface of dissolved particles and with them fall in the bottom depositions. PCBP are stable with respect to most chemical reactions. Their long influence on hydrobionts leads to biochemical and histologic changes, functional violations and violations in the reproduction. The chronic influence of PCBP is very dangerous. Their cancerous action is also noticed.

The concentration of PCBP should not exceed 0.5 ng/l in unpolluted fresh aquatic basin, and in moderately polluted waters and firths - 50 ng/l.

The permissible entry per day of DDT in man's organism should not exceed 0.3 mg (or 0.005 mg per 1 kg net weight). Different kinds of toxic-dermatitis are observed for PCBP in daily use over 0.07 mg per 1 kg net weight.

The presence of chlorinated biphenol is typical for all constituents of Black Sea ecosystems (priming, plankton, crustacean, fishes, weeds, etc.). Highest concentrations of PCBP are found in coastal regions and in river outfalls. The content of PCBP in polychems, echinodermata, crustacean, fishes, etc. is in the limits of 3280-16400 ng/g. The content of PCBP in hydrobionts and sediments from Black Sea changes in large limits.

PCBP content in plankton was between 686 and 2700 ng/g for different depths in 1987 in the regions of Crimea and Bosporus; from 18 to 43 mkg/g in the outfall of the Danube (14 m.depth) and from 130 to 160 ng/g in the central-west region (1000 m depth).

PCBP content in priming is from 94 to 4800 ng/g in the outfall of Bug, about 14000 ng/g in the region of Kerch, 624 ng/g in the region of Evpatoria, to 612 ng/g in the region of Belgorod - Dniester, 427 ng/g in the Bosporus region, 66 ng/g in the central-west region at a depth of 690 m and 1800 ng/g at the coasts of Caucasus. It was noticed that PCBP concentration in bottom sediments in the region of river outfalls has considerable seasonable changes. For example, PCBP concentration in surface layer in the north-west part of Black Sea is 216 ng/g in September and 11 mkg/g in May.

Mussels, as biofilters containing a considerable quantity of lipids, are accumulators of chlorine organic pesticides.

PCBP accumulation in mussels strongly depends on the regions of inhabitancy. In regions away from towns, PCBP concentrations in mussels are low - to 100 ng/g; about 200 ng/g close to industrial regions, to 980 ng/g in strongly polluted regions; 282 and 352 ng/g in the clean waters of Karadag, Laopi; from 493 to 14815 ng/g in the polluted Sevastopol Bay; 2690 ng/g in Fedosian Bay; 1600 ng/g in Kolomian Bay.

PCBP content in fishes and weeds depends on the pollution of their inhabitancy: mackerel, scad - from 450 to 660 ng/g, grey mullet, mugil saliens - from 240 to 560 ng/g, sheat-fish- about 54 ng/g, turbot - 105 ng/g, karas - 96 ng/g, kind of small Black Sea fish - 232 ng/g, weeds-cistoziri - from 556 to 750 ng/g, enteromorph and gymmammodytes cicerellus - from 48 to 78 ng/g, cladocera- 222 ng/g.

To make a quantitative estimation of the assimilation capacity of radioactive contaminations, the concept "radio-capacity" of sea environment is used. By integral approach, radio-capacity is the quantity of radionuclides contained in water as well as in alive and lifeless components. By differential approach, the measure of radio-capacity is the flux of radionuclides eliminated by the considered volume of sea water as a result of radioactive decay and under the influence of other abiotic and biotic factors. The integral and differential estimations obtained in maximally permissible concentrations of radionuclides in water are called radio-capacities.

The interaction between hydrobionts and radioactive isotopes is studied in a constantly entering contamination flow through the water surface layer, on different levels of concentration capacity of the ecosystem components. The profile of vertical distribution of radionuclides has the form of exponential curves. The coefficients of radionuclide accumulation from sea organisms of different systematic groups are between 100 and 10000 units, reaching to one million units for the isotopes of zinc and iron. The specific biomass of hydrobionts for most aquatic basin of Black Sea is from 1 to 10 g/cub. m, reaching to 100 g/cub.m. The biological component of integral radio-capacity usually does not exceed 10 %, but may reach to 90 % of sea environmental radio-capacity.

The anthropogenic contamination increases the intensity of mutation processes and may play the role of an efficient factor for selection, thus furthering adaptation of populations toward inhabitancy conditions. On definite levels of contamination, the adaptive possibilities of populations may surpass and come to degradation. This level should be considered as a measure for the ecological capacity of the damaging action of contamination. Generally, this capacity should be defined by the most sensitive ecosystem link.

Oxidized deep waters do not suppress the vital exchange processes of the weed gymnammodytes cicerellus. The long growing of hydrobionts in oxidized water exerts mutagenic influence on sea organisms in the recreation zone of Black Sea.

Bacteria performing chemosynthesis plays an important role in productivity and biogeochemical processes taking place in basin. Chemosynthesis and its influence reach greatest dimensions in Black Sea. Bacteria performing chemosynthesis are autotrophic. They do not use sun energy but the energy of numerous restored compounds - hydrogen sulphide, ammonia, etc., which appear in depth. The most favourable conditions for development of bacteria performing chemosynthesis are in and about layers of simultaneous existence of oxygen and hydrogen sulphide (C- layer).

The vertical distribution of general and thio-chemosynthesis is of layer character.

The maximum activity of micro-aerophyll chemotrophic bacteria is in C-layers. Under the boundary of oxygen penetration lies the layer of anaerobic chemosynthesis and its maximum. The configuration of this layer is complicated and spreads to a depth of 400 - 700 m. The activity of thio-chemosynthesis bacteria appears sometimes over the C-layer as well. Elementary sulphur may oxidize biologically in the anasulphide layer of chemosynthesis, as sulphuric acid is obtained. The maximal values of val-productivity in the upper anasulphide layer are between 0.00-14.6 mg C/cub.m (average 5.1 mg C/cub. m), from 3.5 to 19.7 mg C/cub.m (average 10.2 mg C/cub.m) in the redox-zone, from 3.1 to 13.0 mg C/cub.m (average

6.8 mg C/cub.m) in the anaerobic layer per day. The maximal velocities of biological oxidization of reduced forms of sulphur are from 38.4 to 623.5 mgS/cub.m per day. The anaerobic layer of chemosynthesis is represented mainly by thio-denitrificators. The micro-aerophyll thio-bacteria give 20-30% of the organic carbon production in the redox-zone. The analysis of the vertical distribution of nitrate concentration shows that nitrite producing bacteria play a basic role.

The general production of organic carbon in the chemosynthesis process is from 0.23 to 1.26 g C/sq.m per day. These values are comparable with the phytoplankton production in Black Sea. Hence the C-layer is a powerful source of organic carbon considerably contributing to the bioproductivity of Black Sea.

The biological oxidization of hydrogen sulphide and its derivatives proceeds simultaneously with the chemical process not only in the C-layer but in the oxygen and hydrogen sulphide zone as well. The C-layer is only a component of the common chemoclin or the biochemical barrier in Black Sea that prevents raising the upper boundary of hydrogen sulphide in upper sea layers, retaining hydrogen sulphide about the boundary of density jump in chaloclin.

The purple and green sulphuric photosynthesizing bacteria are two groups of anaerobic phototrophic bacteria capable to oxidize hydrogen sulphide and other recovered compounds of sulphur in anaerobic surroundings in the presence of light. The vertical distribution of bacteriochlorophyll- Å, typical for the most shadow sustainable varieties of green sulphur-bacteria, has a maximum below the C- layer down boundary.

The influence of anthropogenic pollution on the vertical distribution of different types of biological processes and the relations among them is an important task for fundamental research. As a result, it may be expected:

1. Elaboration of integral and differential criteria for the power of anthropogenic influence.

2. Specifying the estimations of ecosystem self-cleaning capacity;

3. Elaboration of scientifically grounded norms for permissible pollutions by components and seasons, as well as the permissible limit concentrations of different pollutants.

4. Discovery of new unique biotechnologies.

Biological Process between Oxygen and Hydrogen Sulphide Layer Boundary

The strong density stratification of water body defines stagnation and stable hydrogen sulphide pollution of deep waters in open sea. The capacity of hydrogen sulphide layer takes more than 90% of the water column - 1700-1900 m at mean depth - 1900-2000 m. The action of sulphate reducing bacteria is accepted as a basic source of hydrogen sulphide in Black Sea.

The upper aerobic layer is comparatively thin - from 70 to 200 m in dependence of the stream character in the point of contact and season. Specific and unique physical, chemical and biological processes are running at the boundary between oxygen and hydrogen sulphide zones.

The boundary between aerobic and anaerobic zones includes layers of joint co-existence of oxygen and hydrogen sulphide (C-layer, redox-zone) and the subanaerobic layer lying over it with content of dissolved oxygen from 0.4 to 0.2 ml/l.

In layers of joint co-existence of O_2 and H_2S (with thickness from 10 to 80 m), there exist conditions for development of autotrophic microflora in the subanaerobic layer. First of all the chemoautotrophic bacterial production carries out on account of the restored compound energy of sulphur and nitrogen. The newly synthesized bacterial biomass in such conditions may contain a lot of pollutions coming from the oxygen layer in the process of deposition or deep outfalls.

Autotrophic Bacteria and Synthesis of Organic Substance at the Boundary between Aerobic and Anaerobic Zones

According to contemporary conceptions, up to 40% of the total bacterioplankton biomass in the upper 200-meter sea layer is formed by chemosynthesis micro-organisms, connected with the boundary between aerobic and subanaerobic zones. The autotrophic assimilation of carbonic acid in this pelagian field reaches 80% of the total.

The group of chemosynthetic autotrophic thio-microflora defines the oxidization of hydrogen sulphide in the second stage - from thio-sulphate to sulphate. The first stage from H_2S to thio-sulphate passes in an inorganic chemical way. At oxygen deficiency in the lower C-layer part (the layer of joint co-existence of O_2 and H_2S), the oxidization of H_2S possibly occurs on account of thio-denitrification by using oxygen from nitrates. The region of ionic chemosynthesis covers the C-layer and the lower part of O_2 layer (5-15m). Thio-chemosynthesis runs most intensively in the C-layer and its maximum is a little under the upper boundary of H_2S at redox-potential Eh 10-30 mV and concentrations of O_2 and H_2S ~ 0.1 and 0.2 mg/l respectively. This is confirmed by the analyses of bacterial production and microscopic investigations of thread-like bacterial forms, responsible for the thio-chemosynthesis. According to data of Y.I. Sorokin, the production of thio-chemosynthesis in the C-layer is from 40 to 60 mg/cub.m humid biomass daily. More detailed measurements give close values of thio-bacteria production from 24 to 82 mg C/sq.m per day and confirm the localization of thio-chemosynthesis in C-layers and in a very narrow layer over its upper boundary.

Other important chemosynthesis bacteria, inhabiting the region of division between O_2 and H_2S, are the nitrificators creating organic substance on account of ammonium and nitrites oxidization. Estimations on the production of nitrificators in Black Sea have been obtained only lately. It is comparable with the reduction of thio-bacteria and varies in the limits of 25-115 mg C/sq.m per day, that makes up to 60% of the darkish assimilation of carbonic acid. The nitrification maximum is on the surface of the nitrates maximum and in the water layer where an abrupt increase of the NH content is observed. It is connected with the boundary region of H_2S. It is typical for the vertical distribution of nitrification that unlike chemosynthesis, its sufficiently high level is also noticed over the upper boundary of HS - in the subanaerobic layer, and in some cases in oxiclin.

The estimations of the biological productivity of methane-oxidizing (methane-trophic) bacteria in Black Sea differ considerably by different authors. Yu.I. Sorokin points out the very active methane-oxidizing microflora in the layer from the upper boundary of H_2S to horizons of 150-200m. According to recent estimations by D.I. Nesterova, the contribution of methane- trophic bacteria to chemosynthesis production is several orders lower than other chemosynthetics and in total 1.5 mkgC/cub.m per day. The maximal methane oxidization intensity is noted in the C-layer and adjoining water layers.

Still another important group of bacteria, functioning in the upper boundary of HS, are the photosynthesizing bacteria oxidizing the hydrogen sulphide. The action of these bacteria is typical for mero- mectic basin, where the upper HS boundary is in the zone of sufficient light. The works of H. Yanosh and Yu.I. Sorokin in the 70's did not discover this group of bacteria in Black Sea in the upper hydrogen sulphide boundary level at a depth of 110-130 m in this period.

In the spring of 1988, in conditions when the upper boundary of hydrogen sulphide raised to a horizon of 70-80 m, the American expedition of ship "Knor" registered a very high concentration of bacteriochlorophyll "e" in the upper redox-zone part and immediately over its upper boundary. As the analysis of pigment showed, this has been connected with the mass development of photosynthesizing sulphuric bacteria of Chlorobium type. The quantity of bacteriochlorophyll reaches 215-950 ng/l, which exceeds twice chlorophyll "a" concentration in the euphotic layer. The estimations of sulphuric photobacteria biomass give values of 0.5 g/sq.m and in the same time the quantity of phytoplankton is 1 g/sq.m. American scientists connect the appearance and mass development of photosynthesizing sulphuric bacteria with the raising of HS upper boundary and its entering the zone of light. Although there are no estimations on the production of anaerobic bacterial photosynthesis at present, it may play a significant role in carbon cycle and hydrogen sulphide oxidization of sea.

Bringing to an end the examination of autotrophic bacterial population in the boundary area between aerobic and anaerobic zones, it should be underlined that the total production forming there from chemosynthesis is comparable with the phytoplankton production in the euphotic layer. In autumn, when phytoplankton development reaches high level, the mean estimations of primary production are 200-300 mg C/sq.m per day and the total chemosynthesis production changes in the limits of 50-190 mgC/sq.m per day. The specific role of bacterial chemosynthesis in organic substance creation increases in summer and autumn.

Are there mechanisms in Black Sea ecosystem to involve the autotrophic bacterial production, forming at the boundary between aerobic and anaerobic zones, in the trophic chains of plankton communities? The answer may be received after a detailed analysis of spatial distribution of organisms, potentially able to consume chemosynthetic bacterial production (heterotrophic flagellate, infusoria and multicellular filtering zooplankton), on the basis of chemosynthesis vertical distribution.

The structure of hydrochemical fields and vertical distribution of plankton allow to separate three biofluxes in the epi- pelagian of Black Sea: an oxygen-rich zone, including area of sharp O_2 decrease dissolved in depth (oxiclin) with lower boundary at oxygen concentrations 0.4-0.25 ml/l; subanaerobic zone under it of O_2 mean content from 0.4 to 0.2 ml/l and a zone of joint existence of O_2 and H_2S (C-layer).

A characteristic peculiarity of Black Sea plankton vertical distribution in the oxygen-rich zone is the existence of an unusual "false bottom" on the level of lower oxiclin boundary, preventing penetration of aerobic organisms in the subanaerobic layer. The cause for this boundary formation is the O_2 concentration decrease to limiting values 0.2-0.6 ml/l for different groups of plankton organisms. The sea water density increase and the presence of narrow areas with big vertical gradients in the lower pinoclin part may be of importance. Numerous plankton organisms from the aerobic zone, like flagellate, infusoria, phyllopoda, medusas Pleurobrachia, hetognats, compose heapings of density greater than in surface layers on the "false bottom". The difference is from several to hundred times. They spread vertically from several meters to twenty-thirty meters. The concentration of animals in heapings is proportional to the vertical gradients of dissolved oxygen and to water density in the lower oxiclin part.

At C-layer boundary, the plankton population is represented exceptionally by primary organisms - infusoria and heterotrophic flagellate, as a part of them may exist at traceable quantities of hydrogen sulphide. They form clearly localized maxima along the lower vertical part in the subanaerobic layer where the concentration of organisms is very big: flagellate from class Prymnesiida - to 2000 sample/ml, 70 mg/m^3, from Bodonida class - to 20 mg/m^3; infusoria: Pleuronema marinum - to 40-50 sample/l, and from Traheliidae family - to 50 sample/l.

For the present moment three types flagellate and four types infusoria are described in microplankton communities at lower layers in the subanaerobic zone. It should be especially underlined that this association and the community of oxygen-rich layer have not even one common form.

The plankton population of C-layer, besides the abundant and various microflora, includes Protozoa as well. Recently a description has been done by M.V. Zubkov of the specific and very interesting fauna of flagellate and infusoria, inhabiting the C-layer and adapted to existence at significant H_2S concentration. For the present moment two types infusoria and one type of class flagellate are known in the C-layer. The latter meets at H_2S content up to 0.6-1.0 ml/l in concentrations to 100 sample/l. Infusoria of Askinasia class (family Mesodiniidae) is observed at H_2S concentrations from 0 to 0.15 ml/l. Its number reaches 140 sample/l. Mostly massive is the infusoria of class Scoticociliatida, which exists in the presence of hydrogen sulphide and carries symbiotic bacteria. The maximum of its distribution in vertical coincides with the maximum of thio-chemosynthesis, and the concentration reaches 1200 sample/l. The described picture of spatial structure of plankton life in lower layers of the aerobic zone and C-layer allows to estimate, in principle, the possibility for consumption of bacterial chemosynthesis production on account of the restored compounds of sulphur from the population of different biotops.

The field of vertical distribution of fibre forms of bacteria, perhaps "responsible" for thio-chemosynthesis, does not overlap with plankton accumulation in the lower part of oxiclin neither in stable summer stratification nor during intensive winter-spring convection. The direct measurements of bacterial production show that the field of thio-chemosynthesis over the upper boundary of H_2S is no more than 10-20 m. The quantity of thio-bacteria decreases abruptly as well as chemosynthesis intensity higher than the upper C-layer boundary. In some cases, thio-chemosynthesis is not observed at all over the C-layer. This allows speaking about the absence of some considerable direct use of thio-bacteria from plankton forms, inhabiting oxygen-rich waters and forming heapings up in the lower oxiclin part.

Thio-chemosynthesis production may be used by specific nano- and microplankton population of the subanaerobic layer, forming heapings immediately over the C-zone. A similar situation is known for analogous communities of hydrogen sulphide contaminated bottom sediments. However, the investigation of fauna does not show bacterial feeding of dominating flagellate organisms in the subanaerobic layer. This form, on its turn, is a basic object of feeding for a part of the subanaerobic layer, mainly toward bacterial feeding. Besides, the part of thio-microflora in the comparatively poor bacterioplankton subanaerobic layer is not more than 20% in number and 50% in biomass, which certifies that thio-bacteria are of small importance as a feeding object.

The quantity of specific microzooplankton in lower layers of the subanaerobic zone is not big, the spreading of separate forms is characterized by a firm structure with narrow maxima of concentration crossing in vertical. The field of vertical distribution of infusoria and flagellate organisms in the subanaerobic layer does not intersect with the heaping of zooplankton in the lower oxiclin part. All this allows to suppose the absence of some efficient biological "lift", ensuring production transport from thio-chemosynthesis toward zooplankton heaping in the lower oxiclin part by microplankton community from the subanaerobic zone, and first of all toward the migrating phyllopoda-filters. Just this component of plankton may be connected with thio-chemosynthesis production transfer toward the steady links of biological community in the aerobic zone.

The vertical water exchange cannot ensure significant thio-bacteria transport of C-layer through upper subanaerobic layer to the lower oxiclin horizons. This part of aquatic column is characterized by comparatively high vertical stability. The presence of constant structures in the vertical distribution of plankton forms in the subanaerobic layer, the existence of sharp extremes in the structure of hydrochemical fields (of nitrites in particular) confirms the conception for water vertical stability in this field of pelagian.

The given results show that thio-chemosynthesis production, comparable to the level of autotrophic phytoplankton production, is not any natural feeding source for the population in the oxygen-rich zone. This is defined by the oxygen field peculiarities, namely, the presence of a powerful subanaerobic layer separating the field of thio-chemosynthesis and the lower part of oxiclin, where heapings are formed of the most massive potential consumers of thio-microflora. In this way, the organic substance, newly - synthesized from thio-microflora, "closes" in the C-layer community and to some extent in the subanaerobic layer. The field of thio-bios in Black Sea, occupied by a population depending more or less on thio-bacteria production, is limited by C-layers and the lower subanaerobic layer part.

All said about the role of thio-bacteria in pelagian communities in the open part of Black Sea, at the hydrology-hydrochemical structure now existing, is true and concerns also the production of photosynthesizing sulphuric bacteria.

The first estimations of chemosynthesis of recovered nitrogen compounds, obtained recently for Black Sea pelagian, allow to suppose that nitrification as a source of the newly synthesized organic substance for communities in the oxygen zone, as a whole may be of greater importance than thio-chemosynthesis. The production of nitrificators may exceed the production of thio-, and rather significantly the field of their distribution in pelagian is very wider. It includes not only the whole subanaerobic layer but the lower part in the oxygen-rich zone as well. Obviously, this allows utilisation of nitrificators production through the wide presentation of plankton organisms.

How can biological processes change around the boundary between aerobic and anaerobic zone by changing the vertical hydrology-hydrochemical structure of Black Sea?

Obviously, raising the H_2S upper boundary to the zone of light, sufficient for photosynthesizing sulphuric bacteria activity, will bring to an intensive development of this group of micro-organisms. Improving the light conditions of H_2S upper boundary and widening the biotop area are the necessary conditions, first of all in the centers of chalistases that may convert sulphuric bacteria into a very essential source of newly synthesized organisms in pelagian. Respectively, their role increases in hydrogen sulphide oxidization at the upper C-zone boundary.

The cardinal change of the biological processes in the boundary field between aerobic and subanaerobic zones may be a consequence of the intensive vertical convection or raising the basic pinoclin, as the subanaerobic layer falls into the zone of action of synoptic processes and destroys. In principle, the absence of a subanaerobic "silencer" will make the spatial structure of pelagian community in Black Sea, at the boundary between aerobic and anaerobic zone, similar to the structure of communities in meromectic lakes and bottom sediments polluted with hydrogen sulphide. The chemosynthesis production and phototrophic sulphuric bacteria in these communities are a feeding source for the wide spectrum of organisms. In a sharp shrinkage or destruction of the subanaerobic layer in the field of thiobis, bottom heapings of "aerobic" plankton will take part. At that, the autotrophic bacterial production formed at the H_2S upper boundary increases essentially its importance for the community.

It may be considered for the present moment that the vertical dimension decrease of the subanaerobic layer, even to 10 m, will not bring to essential consequences. The hydrochemical structure of water, vertical distribution and character of the interrelations among organisms in the field, including the lower oxiclin part, the subanaerobic layer and C- layer, seem stable compared to the vertical spatial changes of the upper boundary H_2S position from 80 to 200 m.

Development of Ctenophore Mnemiopsis Leidyi and After-Effects for fish industry

The sharp and sometimes catastrophic changes in biological communities (in ecosystems) can be as a result of pollution and anti-science economic activity as well as of intentional (acclimatization) or accidental settlement of new organisms in the basin. This introduction is especially dangerous for isolated basin of weak ecological valence of autochton types which cannot put under control the active settler. Accordingly, if the surroundings conditions are suitable for the new settler - an element of poly-mixed ocean community and for this reason possessing a high ecological valence, then such a type - especially in the initial period of colonization - may bring to enormous bursts in the number and biomass on account of the decreased number of competitive local types.

Black Sea belongs to such semi-relict basins with communities isolated from Ocean and Mediterranean. That is why the settlers falling into it do not encounter a steady competition from the autochton fauna and develop quickly. Such was the behaviour of Rapana, introduced in Black Sea from the Far East, that in practice destroyed the Black Sea mussel beds in no time. Analogous developing now is one type ctenophore (Mnemiopsis) - an invertebrata predatory animal of dimensions from mm. to several tens cm.

The type Mnemiopsis is endemicity in the Atlantic seacoast of North America where there exist two or three close types. One of them (Mnemiopsis leidyi) inhabits, first of all, salty lagoons and estuaries with saltiness close to this of Black Sea and Sea of Azov. Perhaps, this type has been introduced in

Black Sea with ballast waters of ships sailing from America and Canada in the beginning of the 80's. It was observed for the first time in 1982, but its wide distribution began in 1987 in the north-west part of sea, Bosporus region, bays of Caucasus seacoast. In the spring of 1988 its young were encountered in the north-west and Batumi regions of sea, and single big samples (4-5 cm.) - in central regions. In the summer of 1988 it spread in coastal waters of Caucasus, Crimea, Bulgaria, north-west part of sea. In autumn its biomass in the open regions of sea reached to 1.5 kg/sq.m. In the summer of 1989 the total biomass quantity of Mnemiopsis reached billion/ton in Black Sea, and the Sea of Azov was literally overfilled with it. The biomass of Mnemiopsis increased considerably in the spring of 1990.

Such gigantic concentrations of the predatory Mnemiopsis, consuming different organisms of plankton - from infusoria to larvae of fishes, exerted cardinal influence on the structure of plankton communities and fish reserves in Black Sea.

The structural and functional analysis of the state of plankton communities in the basin of Black Sea was done on the basis of data from complex expeditions of IO AN Russia in 1978- 1990. Samples were gathered and treated by a uniform method of taking plankton probes (including mesoplankton) by a 150-l batometer. Mnemiopsis and medusas aurelia (ordinary Black Sea medusa) were registered by catches with fish-net BR 113/140 (dimension 500 mkm) or with conic fish-nets (inlet diameter 50 cm, dimension 200 mkm or 80 cm - 500 mkm), and by a submarine liveable equipment (POA) "Argus", under diving bell and TV cameras (TV). Corrections of medusas and Mnemiopsis leidyi omissions are taken into account because of the fish-nets dimension compared to direct counting from the board of POA. The data for medusas population according to measurements by POA or by TB are two times more on the average than the quantity defined by catches with fish-net BR. Ctenophore mnemiopsis is caught four to five times less than counting by POA.

By several catches are done in the upper layer from surface to thermoclin, in thermoclin and under thermoclin layer by means of fish-net BR 113/140 or other nets at every station. The water column stratification is estimated by the vertical temperature profile, saltiness and density obtained by STD-drill or bati-thermograph.

The estimation of Mnemiopsis and medusas weight is made by different formulas depending on the dimensions. The carbon content in bodies of Mnemiopsis of different dimensions is defined by the wet burning method after a preliminary separation of chlorides. For animals smaller than 10 mm it is about 0.15 %, 10-45 mm in dimension - 1.1 % and for 45 mm - 0.06 % from wet weight.

To characterize phyto-, micro- and macroplankton, samples are used taken by 150-liter batometer at a depth of 100-150 m.

The estimation of Mnemiopsis biomass and number was done in the entire sea basin in March and April 1988 (15th voyage of NIK "Vityaz"). A 80-mile cross-section was done the same year in September by four stations from Gelendjik, when Mnemiopsis was observed in big quantities at coastal sea regions. Every month, from February to November 1989, Mnemiopsis and Medusa aurelia have been caught in the region of Gelendjik (36 samples). 30 stations were made in the summer of 1989 (June-September) during the 44th voyage of NIK "Dmitri Mendeleev". Mnemiopsis, Medusas and mesozooplankton have been investigated in the region of Gelendjik and Anapa (in the 30-mile zone from coast) in the spring of 1990 (the beginning of April) on NIK "Hydrobiolog" and "Aquanaut" in the Sea of Azov (12 stations), and in May the same year on NIK "Academic Boris Petrov" - 20 stations in the west part of the sea.

The seasonable observations of Mnemiopsis population in the region of Gelendjik showed that its young (shorter than 10 mm) meet practically during the whole year, of minimal quantity in the second part of March-April, as the sharply pronounced peak is in July-August. Then the number of young exceeds 9 000 sample/sq.m. The quantity of full-grown specimens in the population is not great and for this reason the total biomass in April-July is minimal. The population biomass increases and reaches its maximum in October-November, slowly decreasing afterwards till April.

In July-August 1989, the mean biomass (natural moisture) in the deep water part of sea increased from 1.4 kg/sq.m - in the period 20 July - 15 August, to 1.6 kg/sq.m from 25 August to 10 September, and the mean dimension of full-grown specimens increased from 18.2 to 26.5 mm. Meanwhile, the biomass of Mnemiopsis type increased from 2.8 to 3.2 kg/sq.m in coastal regions.

The quantity of young prevails sharply in shallow water regions, particularly in the north-west shelf; the mean dimension and big specimens concentrate in the end of the continental shelf, where is the convergence along the external end of the basic Black Sea stream; big specimens of dimension 45 mm and reaching length of 10-13 cm prevail in central rotations. Mostly Mnemiopsis biomass is in the shelf field of the north-west part of sea (to 4.6 kg/sq.m or 44.4 kcal/sq.m). Biomass is less in deep water regions, particularly in the region of slope, where reaches 2.0- 2.5 kg/sq.m (15-20 kcal/sq.m). It decreases to 1-1.5 kg/sq.m in the central parts. A minimal quantity of biomass is measured at Anatolian seacoast in the region of Cape Sinop and Bosporus, where it does not exceed 0.5 kg/sq.m.

Estimation of the total wet Mnemiopsis mass in Black Sea has been obtained by processing these data with the spline-approximation method for the period 20 July - 1 September 1989, equal to 780 million ton (without the Sea of Azov).

In April 1990, Mnemiopsis biomass in the region Gelendjik - Anapa was already 12 kg/sq.m at mean value of the region about 4 kg/sq.m wet weight. This is more than twice from the mean value of Mnemiopsis biomass, measured (1 cal = 4.19 J.) in the summer of 1989. According to the results of NIK "Boris Petrov" (GEOHI AN, Moscow) in May 1990, the mean Mnemiopsis biomass in the west part of sea reached 3 kg/sq.m, as on the shelf at a depth to 100 m it was about 1 kg/sq.m, and at deep water stations about 3.2 kg/sq.m. The maximal value of settler's biomass registered in two stations in the southwest and Bosporus parts of sea is about 11-12 kg/sq.m wet weight, i.e. close to the values measured for the north-east part of sea. According to these data, Mnemiopsis biomass has continued to increase in 1990.

It is necessary to add that in April 1990 Mnemiopsis was absent in the Sea of Azov, and there was no mesoplankton as well. The biomass of the latter varies at separate stations from 10 to 130 mg/sq.m, i.e. 1000 times less than in Black Sea and 10 less than the poor feeding oligo-trophic regions of the tropical ocean. Obviously, the feeding basis of plankton-phages in Sea of Azov is essentially violated by the penetration and mass development of ctenophore mnemiopsis.

The vertical distribution was strictly limited to the biggest gradients of thermoclin, i.e. from 0 to 15-20 m, by the upper surface mixed heated layer in open sea regions in the summer of 1989. Practically, ctenophore did not penetrate in the lower part of thermoclin, all the more in the cool intermediate layer according to data of catches by BR fish-nets. Its young have been met first of all over thermoclin, while the full-grown specimens dropped deeper down and were numerous in thermoclin (mostly in its upper part).

It was established, during observations in October 1989 by the submarine apparatus "Osmotr", that separate big specimens in the region of Gelendjik drop down under thermoclin and even lie on bottom.

Direct observations by a diving bell in April 1990 showed that while the basic mass of ctenophore concentrates around the upper boundary of thermoclin, quite a number of full-grown specimens penetrate in the lower part of thermoclin and under thermoclin waters to the cool intermediate layer at a depth of 50-70 m. Mnemiopsis does not meet in the nucleus of cool intermediate layer.

According to American and ours observations and private communications of E.G. Arashkevich, E.A. Lukanina and O.G. Reznichenko (IO AN Russia), Mnemiopsis can feed on plankton animals of dimension about hundred microns (infusoria) to 10-15 mm (big Calanuses, Sagittas, larvae of fishes). Perhaps it is also able to consume smaller organisms that are stuck to the mucus of its fins. Mnemiopsis does not use specimens of its kind and of the neighbouring one - Pleurobrachia- as food. The opinions on the possibility of consumption of Noctilucas are different; most scientists have observed the use of Noctilucas as food, but A.V. Dritz (IO AN Russia) communicates on the presence of Noctilucas in the gastro-vascular cavity of small (about 10 mm) ctenophores.

Influence on the Structure of Plankton Communities

In March-April 1988 the biomass of small Mnemiopsis was under 30 g/sq.m wet weight and it could not exert significant influence on the structure of plankton communities. However, in September 1988 i.e. 4 months later, the settler's biomass increased more than 30 times and reached 900 g/sq.m in the eastern part of sea. This led to structural changes of plankton: the concentration of Medusa aurelia, a feeding competitor of Mnemiopsis, decreased 30 times; the biomass of Sagitta - hundred times and the smaller representatives of zooplankton - tens. In the same time the concentration of organisms, biotopically inaccessible for Mnemiopsis (Calanus) or consumed by it (Pleurobrachia), remained on the same level for example. The biomass of organisms from more lower trophic levels decreased several times: phytoplankton, bacteria and protozoa.

In April 1989 the 44th expedition of NIK "Dmitri Mendeleev" confirmed the predicted increase of biomass: to 1.8 kg/sq.m, i.e. twice more than in September 1988 (0.8) kg/sq.m. The predictions for a significant change in the structure of plankton communities in Black Sea under the influence of ctenophore mnemiopsis completely confirmed.

The vertical distribution of Medusas aurelia changed in open sea. Earlier, they inhabited the upper mixed layer of maximal concentration in the thermoclin zone and dropped down in the cool intermediate layer to 50-80 m. Practically in 1989, Medusas were "driven" away from the upper mixed layer and almost in all stations, except several that were met along the Turkish seacoast only under thermoclin layers or to its lower boundary.

If for the period 1978-1988 the mean natural wet biomass of Medusas in Black Sea ecosystem was about 1 kg/sq.m (for the whole sea it is about 400 million tons), then in 1989 it decreased to 0.14 kg/sq.m, i.e. 7 times and reached to 60 million tons. Simultaneously the dimensions of aurelia population changed. Till the summer of 1988 the mean weight was about 60 g and since the summer of 1988 it has decreased to 10-15 g. Due to the strong nutritious competition, Medusas cannot reach big dimensions. The mean dimensions decrease of specimens from aurelia population contributes to the preservation of the kind on account of accelerating the physiological processes, since they are inversely proportional to the dimensions of organisms.

In practice, all spread types of zooplankton with the exception of pleurobrachias can be consumed by Mnemiopsis. The observations of the 44th voyage of NIK "Dmitri Mendeleev" and the complex picture in the region of Gelendjik during the whole 1989 confirmed data on the abrupt change in structure of plankton communities. The zooplankton biomass feeding Mnemiopsis decreased 4.4 times in the summer of 1989 in sea deep water part compared to the summer period of 1978, of Sagittas - almost 30 times. The concentration of zooplankton organisms not consumed by Mnemiopsis, in particular pleurobrachias or biotopically less accessible C.helgolandicus, remained approximately on the same level. The phytoplankton and bacteria biomass did not change significantly, and the concentration of Protozoa proved to be higher in contrast to the results obtained in September 1988. The basic parameters and coefficients necessary for structural and functional analysis of plankton communities are measured or calculated.

The estimations showed that ctenophore population of biomass 1.48 kg/ton (10.7 kcal/m) consume about 800 cal/m nutritious plankton per day, i.e. 40% of the observed biomass. In fact, the reserves of nutritious plankton in the upper mixed layer is considerably stronger.

A basic role in the total plankton biomass of plankton community in the summer of 1989 have had phytoplankton, Mnemiopsis, tiny mesozooplankton and aurelia. Their biomass by carbon was correspondingly 25, 18, 17 and 14% of the total biomass. However, the biggest role in destroying the organic substance played bacteria (67%) and Protozoa (20%). Just these two groups of heterotrophic microplankton most intensively mineralized the organic substance created in photosynthesis.

The investigated plankton communities were in a destructive phase of development: the ratio between primary production and heterotrophic destruction was 0.36. The ratio of primary production to the total biomass was equal to 0.07, i.e. close to the index value, typical for low productive regions of Ocean, while in highly productive waters this value is close to 30%.

Thus, visible changes have taken place from the summer of 1988 in Black Sea ecosystem, especially in the structure of meso- and macroplankton in consequence of the development of ctenophore Mnemiopsis Leidyi. First of all the settlement has led to a sharp decrease of biomass in fish nutritious highly caloric zooplankton and its replace by low caloric jelly-like organisms, a part of which (Noctilucas, Pleurobrachias, Medusas, Mnemiopsis) increased, for example, 4 times - from 10 to 40%. In March-April 1988 a part of the jelly-like organisms (close to 40%) increased on account of the small biomass of still undeveloped mesozooplankton and the rather big biomass of Medusas. For comparison, it should be noted that the part of jelly-like organisms in mesozooplankton is close to 10%, i.e. 2-4 times less in the mesotrophic waters of the Pacific Ocean.

Compared to oceanic, the caloricity of Black Sea zooplankton and the carbon content in it are sharply reduced for this reason: till the biomass explosion of Mnemiopsis the carbon content was about 0.4% from the biomass wet weight in mesotrophic waters of the Pacific Ocean - 6%, i.e. 15

times more. The development of new kind ctenophore has led to a decrease of the carbon content of zooplankton in Black Sea twice and reached 0.2%, i.e. 30 times lower than the oceanic.

The part of jelly-like organisms in Black Sea zooplankton, including in macroplankton, is 72-78% in carbon content and 99% in animal wet weight, while in the Pacific Ocean it is 10- 15%. The enormous mass of jelly-like zooplankton, not fit as animal food, eats up the mesozooplankton, spawn and fish larvae that are fit for fish food. These processes run rashly in the thin superficial layer of 50-70 m thickness.

Such influence on the supply and production of nutritious zooplankton makes Mnemiopsis a serious trophic competitor of industrial plankton-eating fishes, first of all, Engraulidae and Sprattus and is the cause for decrease of their reserves.

In the Sea of Azov, where Mnemiopsis reached a particularly large development in 1989 and its mass surpassed 18 million tons, it completely suppressed the nutritious basis of Engraulidae and Sprattus, and also ate their spawn and young. Instead of the usually caught 50-70 thousand tons of Engraulidae and more than 100 thousand tons of Sprattus, the catches in 1989 were a little more than 100 tons. In 1990 at the end of April, only 112 tons were caught and the content of fats was three times less. ("Izvestiya", 1990, N 119.). The Sprattus reserves are estimated to be 400 times less than the former. At that, the biomass of nutritious zooplankton is less than that in the poor oligo-trophic regions of the tropic ocean.

It is mostly alarming that the biomass of ctenophore mnemiopsis in Black Sea continues to increase and the situation with fish catches in 1991-1992 will be almost so catastrophic as in the Sea of Azov in 1989.

In this way, the development of Mnemiopsis type converted in a factor leading to a more catastrophic influence on the biological communities of pelagian and yield of plankton-eating fishes than other direct forms of anthropogenic influence destroying the former unique ecosystem of Black Sea during the last years.

What will happen further?

It is clear that the birth of a new Black Sea ecosystem will begin after two or three years. The question is whether this transition to a new quality will happen without an increase of toxicity of coastal waters, especially in bays. For the present moment there are no elaborated models of these processes.

One necessary but not sufficient condition of transition, safer for the citizens of the Black Sea towns, is the anthropogenic influence decrease in bays.

The quick building of highly technological purifying stations in towns, full stopping of sea pollution with oil products and mineral fertilisers, pesticides and surface active substances over permissible limits by introducing systems for closed water usage and modern technologies in shipbuilding, nutritious, machine-building, power output industry, agriculture and destruction business, is a laborconsuming, expensive and continuous task. The psychological factor connected with insufficient ecological knowledge and responsibility of authorities and population is essential.

People should become conscious of being a part of Nature, not masters.

4

MONITORING AND MANAGEMENT THE ECOLOGICAL STATE OF SEA

The increased pollution of Black Sea determines the necessity of control, limitations and regulation of polluting substances thrown out in sea, i.e. of ecological standardizing the components of anthropogenic influence. This means to apply scientific conceptions and methods for determination of numerical criteria on the ecological state of sea basin, seacoast and bays.

A complex monitoring of sea is necessary to estimate the unnatural changes in structure and function of sea ecosystems, to elaborate norms for the anthropogenic influence.

The concept of natural object monitoring includes three basic components:

First - a developed surveillance net allowing the object state measurement according to the necessary set of parameters of sufficient time and spatial resolving capacity. The net elements may be contact and distant measuring devices on water, land, ship, air and cosmic basis.

Second - communications summarizing the registered parameters, including telephone, telefax, telegraph, radio transmitter and satellite channels for communication and control.

Third - a Center equipped with powerful computers and corresponding software for collection, processing, analysis of data, distribution of received information and, in particular, of supposed decisions. The main task of the Center is to prepare recommendations for managing decisions.

Complex monitoring of the ecological state of sea surroundings means watching biosphere, diagnosis and prognosis of its state, analysis of the extent of anthropogenic factors influence on environment, finding out and estimating factors and sources of influence, and preparing recommendations for regulation of these influences.

Complex monitoring should give information on:

- present ecological state of sea;
- atmospheric transfer role in sea pollution and influence of polluting substances on hydrophysical and biological processes;
- role of river streams and coastal outfalls in sea pollution;
- biological and ecological influence effects of polluting substances on sea organisms and biocoenoses;
- assimilation capacity of sea as an integral characteristic of sea ecosystem capacity for gathering and utilizing polluting substances by preserving the basic properties of biocoenoses;
- prognoses on possible sea state at different extents of anthropogenic influence.

In Black Sea, as well as in other inner seas, a considerable content of chemically active substances is established in hydrobionts, bottom depositions, water column. Lately, the progressing eutrophication in many regions is a result of the influx anthropogenic part of many polluting substances in a volume comparable or exceeding the natural streams in sea.

The dynamism of sea surroundings, the narrow link between physical, chemical and biological processes, the totality of constantly running physical phenomena define:

transfer of technogenic admixtures from streams and their influence on the most vulnerable ecosystems;

appearance of fields with chronic pollution in collection areas of heterogeneous water masses and in the center of static rotations;

transfer of technogenic admixtures in deeper water layers and their concentration in sea organisms and on the suspended organic substance.

The anthropogenic influence on sea increases the concentration of chemical toxic substances in biota, leads to microbiology pollution of coastal regions, diminishes the biological productivity, provokes progressing eutrophication, red tides, violates the ecosystem stability.

Causes for the ecosystem rearrangement might be temporary variations of physical fields, hydrodynamic processes, climate changes, etc. Specifying the contribution of the natural processes and the anthropogenic influence in the ecosystem changes is one of the complex monitoring tasks.

Complex monitoring divides into ecological and physical.

After processing and analysis of data, the ecological monitoring should:

- differentiate the changes of natural character from those caused by the anthropogenic influence;
- estimate the link among changes in the ecosystem with the level of pollutions and make prognoses on them;
- evaluate the critical influence level and most vulnerable links in the biological chain of sea organisms;
- create a measuring system for polluting components and estimating the ecological after-effects;
- work out scientific bases and a system for standardization of the anthropogenic influence on sea ecosystem.

Physical monitoring is a systematic control and analysis of thermohydrodynamic processes of pollution distribution, determining the hydrologic situation and allowing prognoses on ecosystem behaviour.

The estimation of Black Sea ecological situation and its regions, the tendencies for development of the ecological situation are achieved by means of a scientifically grounded analysis - diagnosis of the state of sea, based on historical data and ecological monitoring in the whole sea basin, seacoast and separate regions.

It is necessary for this purpose:

- to prepare annual reports-diagnosis on the ecological state of sea and its regions as well as their publicity;
- to organize complex expedition investigations at least once per season for yield of background characteristics;
- to determine the most vulnerable regions and organize local monitoring;

- to organize satellite monitoring of Black Sea aquatic catchment region by geostation and orbital satellites for control of the hydrometeorological situation and atmospheric transfers.

Regional distant monitoring - observations and analysis of temperature, dynamics, optic-biological and ecological state of sea and neighbouring regions by satellite and sub-satellite equipment making good use of all supplying and accompanying hydrometeorological information.

The control over Black Sea hydrometeorological situation and neighbouring regions, and also over the west, south and north-west and north-east atmospheric transfers, is a necessary element of managing the ecological state and might be realized by geostation satellite "Meteosat" and orbital ISZ "Meteor", "Ocean", etc. Particularly complicated to forecast are Mediterranean sea cyclones, their originating and development, that abruptly change the weather and exert influence, sometimes catastrophically, on the ecological situation.

Satellite Control over the Hydro-Meteorological Situation Atmospheric Transfer and Exchange Atmosphere Sea

The many years experience of satellite information shows that the tendency to synoptic process development in the field of cloudiness is revealed earlier than the appearance of its symptoms in the field of air temperature or atmospheric pressure. The cyclogenesis investigation in atmosphere by analysis of satellite monitoring results in 1986-1989 showed that more than a half of all Mediterranean sea cyclones cross the Black Sea basin and the territory of Ukraine and Crimea. Their formation regions have been determined by cluster analysis. They originate in Ligurian Sea (Genoa cyclones). Typical for them are the prolonged phases of cyclogenesis (-32 h) and fading (-51 h). The mean velocities of cyclones' progressive movement are about 30-35 km/h (maximum to 50-60 km/h). Initially these cyclones move eastward, cross Italy, the Balkans and rush toward the "channel" of width 300-500 km between the mountain chains at the boundary Europe-Asia to Black Sea and neighbouring regions.

These investigations allowed to estimate some cyclones' quantitative characteristics determining the possibility for prognosis on their trajectory and evolution. In principle, this result gives a possibility to estimate the ecological responsibility of industrial pollution of atmosphere over the European continent, North Africa and Near East. The probable estimation accessibility of cyclone's trajectory and its characteristics allows to evaluate the atmospheric transfer contribution to Black Sea pollution. In critical situations (like the Chernobyl accident, for example), the creation of a station informational system will allow immediate measures for control of the situation, in order to soften the ecological after-effects. The spreading of Chernobyl radioactivity showed that an eventual ecological catastrophe in Black Sea region will exert essential influence on the regions of Europe, North Africa and the Near East, lying in the channel of atmospheric cyclone and front exchange.

Informational Software of Sea Investigations

The ecological investigations of sea surroundings have specific requirements toward automation and informational software of the scientific expedition works:

First - a possibility for retrospective analysis of the ecological situation in the region. Namely, the system should include full sub-bases of retrospective data on the measured parameters for the historical period of observations. The system program products have to be adapted to processing and analysis of sub-bases current data, as well as to retrospective sub-bases, for an efficient and operative active diagnosis and prognosis of the ecological state of sea surroundings.

Second - the presence of program products allowing to process and store multi-disciplinary data massifs in quasi real time. Ecological investigations as a rule require measurement of a big quantity of hydrochemical, biological, radiological, hydrophysical and other characteristics and, hence, the automated system has to be sufficiently independent of structure and physical nature of measured parameters, and insensitive to their number. A flexible interactive system is necessary for the control of data bases by easy introducing of parameters and functional possibilities as well. Since the joint influence of several pollutants on the environment is not equal to their arithmetic sum, the system must process and analyze several measurable characteristics simultaneously.

Third - the system must be supplied with mathematical models (diagnostic and prognostic) for the spatial-time unsteadiness in the concentrations of natural and anthropogenic admixtures in sea surroundings, i.e. to have a developed interactive modelling net.

The regular ship's systems for gathering and treatment of experimental data do not ensure complex and operative processing and analysis of characteristics defining the ecological state of sea surroundings, as a prevailing quantity of parameters is measured from samples and is not adjusted to on-line adoption by the ship's systems. Besides, for the present moment there are no standard methods for complex instrument investigations of the ecological state of sea surroundings. A special automated system for gathering, treatment and analysis of experimental data, designed for ecological investigations, has been worked out in the Institute of Geochemistry and Analytical Chemistry in Moscow together with the firm "ECOSY". An original automated relative hydrophysical operative system ARGOS, worked out in the Institute, was used for base of the system.

Main priorities of the system:

- unlimited number of processed parameters (the quantity of sub-bases data is limited only by the external memory volume);
- minimum possible volume of external memory (information is packed in a dual code);
- a developed system of logical criteria for searching data (to hundred simultaneously acting criteria);
- operative access to data (search of any complicated criterion does not exceed several minutes of real time on PC of IBM PC/AT type for data basis containing about a hundred thousand standard hydrological stations);
- interactive working regime of the system in natural language;
- large possibilities for visualization of data (copies on laser or ordinary matrix printer are possible);
- module structure of modelling net (inclusion of new models does not require rearrangement of the common system structure).

Main components of the system:

1) set of sub-bases data differently oriented in problems, disciplines or any other way;
2) a system for control of sub-bases data (program products realizing insertion, fulfilment, reconstruction of data, search of necessary data by different criteria);
3) a packet of applied programs (modelling net) for visualization of initial data, for solving different tasks, including distribution dynamic modelling of different admixtures.

The created system was put in operation on the research ship "Academician Boris Petrov" and revealed the expected efficiency during work in the west part of Black Sea in April-May 1990.

The system ensures work with experimental data in a dialogue regime and in "on-line" entering of data in almost real time. The wide set of functional menu allows to visualize the whole reference and auxiliary information on screen in different forms, as well as the results of data processing and analysis, including the system informational fund, station passport, numerous meanings of the measured parameter, station location in the co-ordinate net, horizontal distribution in iso-lines, vertical distribution in iso-lines, vertical profile of a separate station.

The system is worked out in English.

During the above mentioned voyage of NIK "Academician Boris Petrov", the following results have been obtained by the described system:

1. Formation and processing of the historical massive of hydrological data on Black Sea. Over 55000 stations have been processed covering the end of last century until now. About 12000 stations have been selected out of them referring to the west part of Black Sea and about 500 stations referring to the period from 26 April to 15 May. According to these data, the spatial distributions of hydrological characteristics (temperature and saltiness) were built and analyzed for separate stations interesting for estimation of the anthropogenic influence on surroundings;

2. Based on results from experimental investigations in the north-west part of sea, opposite to Bulgarian seacoast and in Burgos Bay, sub-bases of data have been created in general fields as radiology, biology, hydrology; oriented by disciplines - chemistry 1, chemistry 2, chemistry 3; oriented by regions - Burgos 1, Burgos 2, Burgos 3. The bases include more than 50 parameters.

Horizontal and vertical distributions have been built for different parameters and compared with distributions from the historical massive where it was possible.

5

BURGAS BAY

The firm "ECOSI" was founded in the end of 1989 on initiative of the international scientific staff "Black Sea". The abbreviation "ECOSY" means an express ecological analysis of sea regions for taking natural protection decisions on all levels of management, economy, technology and other accompanying activities. The reason was that our recommendations of May 1989 on control over the ecological situation in Burgos Bay, although accepted by local authorities on 14 of July 1989, practically have not been implemented. We supposed that the organization of a firm will intensify the investigations, elaboration and their implementation in practice, and support to overcome more efficiently the scientific institutional, regional, national and party feudalism in the fulfilment of "Black Sea" program approved at the Seminar "Pomorie 88".

Further on, we shall give the basic results for Burgos Bay obtained in the process of elaboration of Prognosis - a diagnosis for the state of bay with recommendations for management of the ecological situation, worked out mainly on public basis as a social commission by the Burgos state authority from October 1988 to May 1989.

5.1. PHYSICAL AND GEOGRAPHICAL PECULIARITIES OF THE REGION

Burgos Bay occupies the east part of a big depression morphostructure. Geomorphological depression borders in the north are the slopes of Aitos and

Karnobat mountains. To the west it is limited by a chain of low hills. The southwest and south depression borders pass along the north parts of Strandja. Its land part (3785 sq.km) drains by a rather dense river net. Situated from south to north are fresh-water lake Mandrensko and salty Burgasko (27.6 sq.km) and Atanasovsko (16.9 sq.km) lakes, and Pomoriisko lake-lagoon (6.7 sq.km).

The area of Burgos Bay, in the west from the meridian of Pomorie, is 174 sq.km. The ratio of basin aquatic catchment area to bay area is equal to 21.7. This big figure is one of the criteria for the increased risk of anthropogenic influence.

The coastal part relief of land joins hill elements of surface with terraced plains and the biggest Bulgarian firths and sandy beaches.

The shelf part is with flat surface and a slight slope to the east. At a depth of 50 m, the most essential morphological elements are relicts of a submarine system of valleys and Etinski swell, which is formed of sand-alevrit material and has hydrogen genesis. There are several local elevations and gullies in the central part of shelf. The external zone of shelf (under the 100-meter isobar) is a narrow stripe with big slopes at bottom, which turns into a continental slope. The continental slope relief is threaded by different in size submarine valleys, grouped in several branch systems, cutting through slope and proceeding to abyssal plain.

The geological structure of Burgos region has been formed by powerful vulcanogenic and vulcano-depositions complex of types in senon age with implanted magmatic bodies. The various number of types includes trachytes, trachoandesites, andesites, tufts, tuftlets and tuftbreaks, all enriched with heavy and ore minerals. There are polymetal ore fields in Medni Rid. Their processing and beds are in the proximity of the traditional resorts. There are younger types of Paleocene, Neocene and the Quaternary over late melovites in the coastal sea part. There are carbonate clayey types of coal. In Miocene depositions enter shallow water lime sandstone, clayey marls, etc. Pliocene depositions are of poorly cemented conglomerate of sands and clays of lake genesis. Clays prevail in the north parts of region. The Quaternary layers in region are of river-lake and shallow water-sea origin. The maximal capacity of the Quaternary depositions is in Mandrensko lake and reaches 60 m.

The big Burgos Bay bottom is covered mainly by late Quaternary depositions and contemporary depositions. Under a depth of 5-10 m, alevrit and alevrit-pelit components appear and, practically, there are no sand formations under 20 m. Typical for most sand formations is the high content of a heavy magnetic fraction represented by minerals connected with magmatic and volcanic rocks from the continental part.

Clayey-alevrit and lime-alevrit-clayey slimes, green grey and oil green from liquid to soft lithe consistency, are spread in shelf central part in the range of 50 to 80 m. Biodetrite sands have local development. Ancient root types are discovered in places with strong bottom stream.

Summarizing, it may be said that clayey formations of various mineral content prevail. More of them possess good sorption properties and concentrate ions of elements, molecules of organic compounds and other substances of technogenic origin. The behaviour of the latter in natural surroundings, their influence on biota in basin is an object of future complex systematic investigations.

Contemporary depositions are formed from material of "Danube origin" in the north and in west by abrasion and coastal landslidings in the region of Sarafovo. The steady flow entering Black Sea from the Danube is about 87 million tons/per year, while the whole Bulgarian coast under the influence of sea abrasion gives a little more than 1.34 million tons/per year deposition types.

In the last years, due to the participation of Venelin Velev, geologist, in expeditions of NIK "Academician Boris Petrov" in 1989 and 1990, a survey-inventory scheme has been drawn up of submarine valleys; hierarchy of river beds and tributaries were evaluated, some qualitative and quantitative indexes of Quaternary depositions on the continental slope and bottom in front of Burgos Bay were defined. Relief morphology testifies about high lyto-dynamic activity in the region occurring on shelf, as well as on the continental slope.

5.2. HYDROMETEOROLOGIC - REGIME OF BURGAS BAY

The main factors determining the hydrological regime of bay are: coast configuration and bottom relief, regional meteoconditions, water exchange with the west part of Black Sea and rivers tributary.

West and north-east winds prevail in autumn and winter, while west, south-east and east winds in spring and summer. Typical for summer are local breeze winds.

According to data of the satellite system "Landsat", a cyclone circulation is possible at weak wind in bay, at coming out of bay's waters along the south coast. At normal weather, a circulation is possible with two streams in the southwest part of bay: to north-east and the second one - through the center of bay in open sea. Streams velocities in the north part of bay are bigger than these in the south. Typical peculiarities are local whirlwinds in bays in the south part.

The decrease of depth at bay entrance is the cause for breaking the wind waves in winter, which has a positive ecological importance because it airs the bay.

The thermohalin structure of waters from May to October has a well pronounced density stratification with maximum in July-August. The saltiness of surface layer is 0.03 parts pro million less than that at bottom. The temperature difference between surface and bottom layers in the period from May to October may reach to 6-8 degrees C.

5.3. STREAMS IN BURGAS BAY

Data from cosmic observations for water's circulation in bay are confirmed from nature measurements. Velocities from 0.02 to 0.16 m/sec were measured in different horizons. Velocities are highest at a depth of 2 m. Water masses movement in bay is first of all from west to east and from west to north-east. The main cyclone stream is traced from the south-west coast and has several whirlwinds, whose arms enclose the bay basin. The closed character of streams creates conditions for retaining the polluting substances in bay. The minimal time is about 5-6 days and the real time is more than 18-20 days.

Four zones with complicated hydrodynamics outline in Burgos Bay, namely in the region of Cape Chukalya - the isle of St. Anastasia, Cape Atia - Cape Akin, town of Pomorie - Cape Lahna. Polluting substances fallen in bay retain and accumulate in these regions.

The solutions of hydrodynamic equations in approximation of shallow sea at different wind situations co-ordinate with the investigations from nature specifying them in detail. They show:

existence of basic half-closed circulation with streams acceleration in the
    north part and sufficiently stable orographic whirlwinds in the south;
anti-cyclone main circulation and cyclonic orographic whirlwinds at west;
the main circulation type is cyclone and the orographic whirlwinds - anti-
    cyclone at east winds.

The wind change at normal weather with night and day breezes brings to a change in the circulation type after several hours.

At wind velocity 10 m/sec, the streams reach velocity to 35 cm/sec in the north part.

5.4. MODELING THE OIL PRODUCTS DISTRIBUTION

One of the main tasks of applied ecology of sea regions is the estimation of assimilative capacity of every single pollutant and their totality, and time minimization for express diagnostics of the ecological state.

The assimilation capacity characterizes numerically the permissible limit anthropogenic influences that do not bring to irreversible changes in the basin ecosystem. The comparison of this value with quantity of different pollutants falling down in basin is a practical criterion for the right solution about the number of purification stations and quality of technologies and systems for purification of wastewaters.

For express diagnostics of the ecological state of bay, first of all there should be scientifically grounded and qualitatively reliable methods for minimization the measuring time of pollution components. One of the ways is a preliminary modelling of hydrodynamics and the distribution and utilisation of pollutants in bay. Our experience in Burgos Bay in 1988 - 1990 has shown that time and expenses might be reduced ten times in this way.

As oil products are the most spread and dangerous pollutants in Burgos Bay, modelling their distribution is an important element in ecological monitoring.

Three numerical experiments have been carried out with different data for the quantity of pollutions in streams corresponding to the prevailing winds (west, south-east, north-east):

1. Determining the assimilative capacity of bay with regard to pollution with oil products. Mass transfer, evaporation, deposition and the totality of chemical and biological transformational processes of oil hydrocarbons were reported;

2. The mean oil concentration in surface layer was calculated by the number of known polluting sources of oil products (2.6 tons/per day according to official data, distributed as follows: Pomorie - 9.5 kg/per day, Sarafovo - 1.3 kg/per day, Kraimorie - 17.3 kg/per day, Burgos - 1146 kg/per day, NHK - 1445 kg/per day. It proved to be equal to 0.02 mg/l at 0.05 mg/l permissible limit concentration (PLC). The comparison of mean concentration value and spatial picture of spreading in bay with the experimentally measured showed that official data were strongly diminished because they do not record all polluting sources of oil products from harbour, oil harbour, local enterprises of machine-building and transport in Atia, Kraimorie, Burgos, Pomorie, Nesebar and Sunny Beach.

3. Recording the real pollution, including oil flowing in the harbour (9000 kg/per day) from ships (1500 kg/per day), the mean concentration of oil products equal to 0.19 mg/l or 4 PLC for instance has been obtained which already co-ordinated with data from nature. The maximal value of oil products concentration, measured in bay, is about 11 mg/l in the region of tanker limbering.

Analysis of the scenario at different wind situations allows to divide the bay into regions by the extent of risk for oil-contamination from own sources.

The acquired experience shows that for an express diagnostics of Burgos Bay pollutions it is sufficient to watch from 3 to 8 points in the zones of maximal ecological risk if the observations are accompanied by modelling in real time.

5.5. HYDROCHEMICAL AND SUMMARIZED INDICES OF WATER QUALITY

The hydrochemical regime of Burgos Bay is considered according to the following parameters: temperature, saltiness, pH, oxidizing recovery potential (Eh), content of nitrites, nitrates, ordinary nitrogen, ordinary and mineral phosphorus, content of phenol and 9 of its derivatives. The content of pesticides (by the sum of cyclodiens) has been measured in different points of bay.

Simultaneously with hydrochemistry investigations, the quality of sea water is determined by the integral toxicological index, according to the extent of survival of glowing bacteria Benecka harveiyi.

The main purpose of investigation is the distribution of polluting substances and also estimation of the capacity of sea aquatic surroundings to recover the normal, biologically complete state. Kinetic approach has been used in estimating the self-cleaning capacity of water. It consists in using the summarized indices of water quality, characterizing the capacity of basin for chemical and biological transformation of fallen pollutants. These indices are kinetic because characterize the dynamics of chemical and microbiological processes running in the basin and leading to self-cleaning.

31 stations have been placed in the standard net based on the hydrochemical and hydrobiologic indices of Burgos Bay. Besides, samples have been taken from 10 shore and coastal points, 6 of which (freshwater) are main sources of pollution, and 4 are in the proximity of shore at beach zones.

Biogenic substances, to which nitrogen and phosphorus compounds refer, are the necessary component of natural aquatic surroundings. In natural waters nitrogen occurs first of all in the form of minerals (NH_4^+, NO_2^-, NO_3^-) and in organic form (amino acids and organic compounds). The increased content of ammonium form of nitrogen points at "fresh pollution" with nitrogen containing substances (everyday wastewaters, waters from stock-breeding farms, nitrogen containing fertilisers washed away from fields). Ammonification of basin takes place during transition from aquatic surroundings into steady recovery state in absence of oxygen. Nitrite - ion participates as an intermediate compound in the row from NH_4^+ to NO_3^-, appearing as a product of nitrate - ion recovery or as a product of oxidization of nitrogen containing compounds, standing in lower states of oxidization: NH_4^+, NH_2OH, N_2O, NO. By nature, nitrite-ion is a strong toxicant.

Phenol and its derivatives are among the most spread components of natural water pollutants. The presence of phenols is connected with the natural processes and the anthropogenic influence as well. Phenol and its chlorine-, nitro- and alkyl-derivatives are often present in wastewaters of industrial enterprises, such as: cokechemical, pharmaceutical, dyes and varnish production, woodworking, paper and cellulose production and so on. Besides, chlorine derivatives of phenol are formed at chlorination of drinking water, as many of them have high toxicity (including cancer).

Recommendable are the following values for permissible limit concentrations of phenol in waters for everyday life: phenol - 0.001 mg/l ; 2,4 - dinitro-phenol - 0.03 mg/l; 2- nitro-phenol - 0.06 mg/l; pentachlorinephenol - 0.3 mg/l; 4- nitro-phenol- 0.001 mg/l.

To identify pollution sources and obtain a full characteristic of environmental pollution, from 7 to 15 phenol derivatives should be determined.

The method for phenol analysis is based on the principle of selective division of mixture from components by means of highly efficient liquid chromatography with identification of different compounds according to retaining time. An UV detector is used for registration of components. The method allows to analyze the content of phenol and its 9 derivatives in water. The range of measured concentrations in mg/l and the absolute error boundary measurement in mg/l is the following: phenol from 0.001 to 0.03, 0.0005; 4-nitro-phenol from 0.02 to 0.6, 0.01; 2- nitro-phenol from 0.06 to 1.8, 0.03; 2-chlorinephenol from 0.30 to 9.0; 0.15, 2,4- dinitro-phenol from 0.03 to 0.9; 0.015, 2,6- dimethylphenol from 0.01 to 0.3, 0.005; 4-chlorine-3-methylphenol from 0.01 to 0.3, 0.005; 2,4- dichlorinephenol from 0.01 to -0.3, 0.005; 2,4,6- trichlorinephenol from 0.3 to 9.0, 0.15; pentachlorinephenol from 0.3 to 9.0, 0.15.

The method was certified in the Research Institute of the Meteorological Service, Moscow. This method may be applied in content analysis of phenol type pollutants in sea water and wastewaters from enterprises.

The method for biotesting sea (and fresh) water by the cease of luminescence in glowing bacteria is Based on the estimation of suspension luminescence intensity from bacteria cells under the action of the investigated polluting substance, or in the investigated water sample compared with the glowing intensity of controlled untreated cells for a given time.

Two more objects have been used for fresh water samples: the small crab Daphnia magna (a current test for chemical pollution, often used for estimation of wastewater's toxicity) and infusoria Tetrahimena pyriformis. This infusoria is sensitive to the change in the oxidizing-recovery state and the presence of toxicants of reduction nature.

The main water hydrochemical indices of Burgos Bay in May 1990 were: water temperature varied in the ranges 11.5 - 13.59°C, pH 8.0 – 8.2; oxygen content 9.0 – 9.9 mg/l. The concentrations of nitrites and mineral phosphorus were close to zero, nitrates from 0.005 to 0.06 mg/l, total nitrogen from 0.05 to 0.28 mg/l, total phosphorus 5.9 – 19.0 mkg/l. The values of these quantities indicate the small contribution of the biological component, typical for that period of the year, and are mean values for Black Sea according to last year's data. However, in summer after water warming over 20°C, these quantities strongly change, especially in the period of micro-weeds blossoming.

All water samples of Burgos Bay along the net of 31 stations were intoxic according to the bacterial biotest. Yet we know that for the correct analysis of the biological toxicity of sea water, one test in May is completely insufficient. Furthermore, by analogy with Odessa Bay, an increased toxicity should be expected several days after the blossoming maximum in summer in comparison with last year. This prediction confirmed in the summer of 1990.

Water samples for analysis of phenol and nine of its derivatives were taken from surface in station points (see the map of Burgos Bay), as well as in points at inflow places of industrial and urban wastewaters, at beaches and places for garbage discharge.

The quantity of phenols in Burgos Bay varies in different points from 0.02 to 0.08 mg/l and the mean value is 0.054 mg/l, which strongly exceeds the value of PLC.

Typical pollutants in this zone are ortho- and para-nitro-phenol, 2,4-dinitro-phenol, 2- chlorinephenol, 4- chlorine, 3- methylphenol, pentachlorinephenol.

The use of program means specially created for these investigations, including computer graphics, gave the possibility to isolate the places with biggest pollution of the bay basin and trace the changes in the investigated compounds concentrations and possible ways of their distribution. Data on the spreading values of all phenol quantities in Burgos Bay completely correlate with streams' direction in west and northwest wind observed in the period of taking the samples.

Using the distribution profile by iso-lines and the concentration vector increase of every single pollutant, the places can be defined from where the pollution of bay comes. The juxtaposition of data on phenol spreading in bay and data of water samples analysis from different coastal points in proximity to outfalls allows to identify the pollution source more precisely, as every source characterizes with its own set of prior pollutants. Such a program cannot be fulfilled with sporadic ship's investigations. For the purpose, it is necessary to set the first part of the laboratory in Pomorie, according to decisions taken in the Seminar "Pomorie 89" and the Executive Committee decision of the District Council in Burgos of 14 July 1989.

Thus, for example, nitro-phenols come in basin from bay bottom discharged in great quantities with wastewaters of NHK. Their quantity in wastewaters of NHK is: 4-nitro-phenol - 0.049 mg/l, 2,4- dinitro-phenol - 0.4 mg/l, 2- nitro-phenol - 0.192 mg/l. Along with that, an essential contribution to pollution of the harbour part of bay have the deposition loads from harbour complex and railway station (2,4-dinitro-phenol - 0.009 mg/l, 2-nitro-phenol - 0.022 mg/l), and also wastewaters from the old part of Burgos (2,4-dinitro--phenol - 0.036 mg/l). Channel waters in the harbour complex (1.042 mg/l) contain phenol as well as wastewaters of the residential complex "Tolbuhin" (0.467 mg/l), and the rain collector of residential and stock-breeding complexes (0.109 mg/l). Traces of phenol are noticed in the Burgos Bay basin in proximity to the central beach, from harbour toward north-east (along the Bay stream). 2,6- dimethylphenol is noticed in the basin east of Burgos and in water samples taken from piers of town's beach. Here pollution comes from complex "Tolbuhin", where the concentration is 0.027 mg/l.

Pollution in the north-east part of Burgos Bay basin with 4-chlorine 3-methylphenol and 2,4-dichlorinephenol is performed by the bay streams which deliver powerful pollution sources to bay bottom: harbour complex, complex "Tolbuhin", airport in Sarafovo situated near the coastal line.

Proceeding from obtained data, a conclusion can be drawn that the biggest part of phenol entering Burgos Bay is brought in by wastewaters of NHK, harbour complex and railway station, wastewaters from the residential complex "Tolbuhin", and also by the rain collector of stock-breeding and residential complexes to the west of lake Atanasovsko.

In conclusion of the chapter on phenols, we shall note that some phenol derivatives, such as pentachlorinephenol, 4-chlorine, 3-methylphenol, 2,4,6-trichlorinephenol, practically do not participate in the biotic rotation and possess high toxicity with regard to hydrobionts.

Similar, even more dangerous for live water systems, is pollution with pesticides. Their transformations in environment run very slowly. Pesticides accumulate in the fat fractions of aquatic organisms. In the period of insemination when fishes do not eat and spend their fat supplies, they poison and die and offspring is often with defects. The consumption of such fish brings to heavy allergic diseases, stomach and liver diseases. In general, nowadays, pollution with pesticide levels by force of influence on living organisms with radioactive contamination. By preliminary data of doctor Splid (Institute of Ecology, Denmark), working on NIK "Academician Boris Petrov" during the research of Burgos Bay in May 1990, the total content of pesticides (eldrin, dieldrin, chlorinedan, heptachlorine,etc.) in Bay waters is about 50 mkg/l, which exceeds 50 times their content in open sea.

An important property of natural aquatic surroundings is the self-cleaning capacity. The totality of physical, biological and chemical processes are taken into account, as they drive polluting substances beyond ecosystems' boundaries, their transformation and redistribution among different objects in the surroundings. In essence, the self-cleaning of surroundings is achieved only by microbiological and chemical transformation of substances to end intoxic products. The processes of mass transfer and redistribution of polluting substances have a seeming effect of self-cleaning, because increasing the zone of anthropogenic influence, they lead to a decrease in the concentrations. At the same time they pollute the neighbouring part of environment, store polluting substances in bottom depositions, collect them in biota and so on.

The processes of mass transfer of polluting substances into aquatic surroundings can be already sufficiently well parametrized and mathematically modeled. The description of the chemical and biochemical transformation of polluting substances is much more difficult.

Only easily assimilated organic substances yield to microbiological transformations, as well as substances yielding to full, hydrolytic and peroxidized oxidization. The microbiological transformation velocity of polluting substances is proportional to bacteria biomass (number) and current concentration of substance. Due to the velocity constant dependence of biolysis on the physical factors of surroundings (temperature, saltiness, conditions of mixing) as well as on their chemical content (content of co-factors, nutritious substances), the contribution of biological self-cleaning varies on large scales. This contribution was negligible in the west part of sea in May 1989.

At the same time micro-weeds may participate indirectly in self-cleaning of aquatic surroundings. During the photosynthesis processes and influence of the short-wave component of sun's ultraviolet (UV) radiation on cells of weeds, they release hydrogen peroxide outwards. At that, a rather high standing concentration of H_2O_2 - to 10 M is established in the inner cells' surroundings of weeds under the sun light action. Taking into account the circumstance that weeds contain H_2O_2, it is clear that a high efficiency is ensured in the peroxide oxidation of substrates - donors of atoms H (phenols, amines).

It is known that some types of weeds, for example the greenish blue, release substances with reducing properties in the environment. These substances interact with oxygen and hydrogen peroxide as catalysts of metal ions of variable valence dissolved in water, in particular, ions and copper complexes. As a result, free radicals and other products of oxygen and hydrogen peroxide may appear in the aquatic surroundings. Especially active particles are formed at interaction of metal ions in the recovered form with hydrogen peroxide. Thus, the realization of conjugated radical processes of oxidization of hardly oxidized pollutants is possible along with oxidization reductionists of natural origin.

The estimation of these reactions contribution is made by the H_2O_2 rate velocity in water, as the hydrogen peroxide formation as well as its decay are carried out by the intermediate formation of free radicals, in particular, of OH radicals.

The different photochemical processes running in sea surroundings under sun radiation action lead also to formation of free radicals. This is photolysis of chumin substances, nitrates and nitrites, photolysis of metal complexes, in particular, complexes of copper chlorides, complexes of iron with oxi- and fulvo acids, disintegration of H_2O_2 and organic peroxides and so on.

Many of the polluting substances disintegrate into harmless ones under sun UV-irradiation. This occurs due to direct sun radiation action as well as indirectly by formation of electron excited particles - sensitizers and free radicals.

The formation of quasi-reduction conditions influence unfavorably on the ecosystem state (when oxygen is present in water and at the same time the substances efficiently interact with H_2O_2). In particular, the aquatic surroundings self-cleaning capacity decreases under these conditions. Substances titrated with hydrogen peroxide are also found in Bay bottom slime which testifies for the significant unbalance of oxidizing-recovery processes in inner basin. Bottom depositions contain 2.10^{-4} g-eq./kg reduction equivalents titrated with hydrogen peroxide.

For example, we obtain for oil products (PLC 0.05 mg/l) estimation of their permissible discharge in Burgos Bay water A = 700 kg/per day. In summer this value is bigger but not more than several times - to 2000 kg/per day. The contribution of other processes in water self-cleaning, such as surface evaporation, bottom deposition, direct photolysis, biodegradation, might be essential. However, for a better estimation of the last two factors at least, laboratory investigations are necessary for the separate polluting substances in the Bay according to place and season conditions. This is only possible in conditions of well equipped and permanently working laboratory.

The total assimilation capacity of Burgos Bay toward pollutions with bionic substances on account of all radical, peroxidized and catalytic redox-processes of self-cleaning can be estimated by the total hydrogen peroxide mass forming and disintegrating in water. The present estimation is about 50 tons/per day. For substances of poor volatility and typical low sorption capacity in suspended particles and bottom depositions, and recording the low biological self-cleaning efficiency of aquatic surroundings with regard to hardly oxidized pollutants, this value is most probably an estimation for maximum permissible discharges of such substances in Burgos Bay waters.

In the case of easily assimilated substances whose self-cleaning proceeds first of all through the biological processes (feeding), the permissible limit discharges should not exceed the total primary production.

Water samples have been investigated separately from 10 coastal points, 6 of them from wastewaters of enterprises and residential complexes of Burgos, and 4 - from beach zones of Bay. There are analogous data for these points of

May 1989 (Seminar "Pomorie 89"). The toxicological control of freshwater samples has been carried out by means of three biotests:

1. Embarrassed vitality of glowing bacteria - determined by comparison of bacteria glow intensity in the sample and in the controlled surroundings;
2. Investigation of population reproduction of infusoria Tetrahimena pyriformis - sharp toxicity characterizes with survival of infusoria for 15 min, chronic - in the course of 96 hours;
3. Population reproduction of small crab Daphnia magna in the course of 96 hours.

In case of sample toxicity, the extent of diluting is determined for toxicity removal - coefficient of toxicity.

It should be noticed that samples from NHK waters do not show toxicity after the biological lakes, although the value of HPK is significant. This testifies for the efficiency of biological lakes in disintegrating some toxic substances in wastewaters of NHK. High toxicity was measured in the analysis of unclean waters, discharged in the harbour complex by the Pilot Service. The sample characterizes with a low redox-potential and a considerable content of substances of reducing type, titratable with hydrogen peroxide. A decrease of toxicity was achieved to zero for infusoria Tetrachimena pyriformis by sample hundred times diluting.

The reduction conditions are still stronger expressed in wastewaters from the old part of Burgos without cleaning. Here sharp toxicity is observed in the small crab Daphnia magna. Substances possessing strong oxidizing properties have been registered instead of reducers in water sample taken from the channel connecting lake Vaya with sea at "Fish Industry".

The most sensitive estimation biotest of water redox-state are glowing bacteria. This allows to recommend the test for express estimation of the negative influence of discharged wastewaters on ecosystems by number estimation of the oxidizing-recovery processes in basin.

Comparison of data for phenol toxicity and content shows that the toxicity of analyzed wastewaters does not correlate with the phenol content and its derivatives in them. A much more essential factor is the oxidizing-reduction state of discharged waters. At the same time, recording the capacity of hydrophobic substances for bioconcentration and the phenol capacity and its derivatives for oxidizing-recovery transformations, it may be said that the pollution of Burgos Bay aquatic surroundings with these substances, whose concentration exceeds many times the PLC, leads to serious structural changes in Bay ecosystem.

The radioactivity of shelf's waters in the west part of sea is increased due obviously to technogenic causes: open unloading of ores and fertilisers in harbour and beds of ore-dressing factory in Meden rudnik. There are also cobalt Co-90 in some places. The mean content of uranium U-238 in surface waters of the Bulgarian seacoast is 2.0 g/cub.m and reaches 2.8 g/cub.m in some points of Burgos Bay.

Radioactivity of Vromos Bay

The radioactive contamination of beach and bay is due to many years discharge of wastes from mine Rosen. The sand gamma-activity from the beach in the discharge region varies from 560 Bq/kg to 17800 Bq/kg only for the isotope of radium Ra-226. The soil is at least forty times more contaminated in the region of mine and bay than the mean value for coastal soils. The bay bottom is contaminated in the limits of 1400 Bq/kg. The norm of sand used in building is up to 370 Bq/kg. That is why taking sand from the bay for building use is absolutely inadmissible without preliminary measurements. The investigations in 1989 showed that the mine still throws wastes in the bay.

The sand layer in the bay is of capacity to four meters and according to the extent and depth of contamination is divided in three zones: strongly contaminated with area of 15 sq.km, radiation over 370 Bq/kg and capacity two meters; average contaminated area of 21 sq.km, radiation from 190 to 370 Bq/kg and capacity one meter; the rest part of radiation sandy strip to 190 Bq/kg and depth to thirty cm. It is high time to encircle the beach and put corresponding informational panel warning people and especially families with small children going to bathe. Some of the private estates are contaminated and farming is not recommendable without preliminary measurements and eventual cleaning. The level of subterranean waters is from

1.5 to 2 m. Well water in private estates is not contaminated and may be used for irrigation.

5.6. PHYTOCENE STRUCTURE AND WATER-TROPHIC IN BURGAS BAY

The high values of phytoplankton biomass, chlorophyll and primary production along the west seacoast and especially in bays deep into land, testify for big water-trophy. The background meanings of the biological parameters in these regions are: phytoplankton biomass - 3-4 g/cub.m (for comparison 100-150 mg/cub.m in open sea), chlorophyll from 0.5 t to 7.5 mg/cub.m, primary production - from 1 to 2 gC/sq.m per day. These indices increase ten to hundred times during "blossoming" of di-atomic and especially of pyridine weeds ("red tide"). In these cases the biomass reaches 1 êg/cub.m, chlorophyll "a" reaches several hundred mg/cub.m, the primary production reaches 20 gC/cub.m per hour.

During blossoming and especially in the period of weeds dying, a great quantity of organic substance is formed. A certain part is utilized by plankton organisms, another part sink in bottom. The organic substance sunken in bottom creates danger for secondary organic pollution and, as a consequence, phenomena and changes in water quality can be measured. Zooplankton utilizes about 5% of the available biomass in the period of "blossoming". This figure is certainly reduced. The investigations in the last two years, Based on luminescence microscopy of "live" phytoplankton, allowed to separate a numerous group of heterotrophic weeds, traditionally referred before to autotrophic. This group can use organic substances as a main source of energy. Heterotrophic weeds are present in practice during the whole year in coastal phytocene and sometimes are more than 50% of the phytoplankton biomass. The role of heterotrophic weeds is comparable with the role of microzooplankton. The strong development of nano- and microzooplankton, of heterotrophic weeds, are sufficiently precise indicators for eutrophication.

The extension of aquatic basin seized by "red tide" and increased frequency, especially after 1986, had led to a qualitative change of phytocenosis, manifested in the contribution increase of peridines in the number and biomass of phytoplankton. One of the probable causes for such a change might be the use of suspended and dissolved organic substance from peridine weeds. Thus, in the time of decreased "blossoming" of Exuviaella cordata, a transition is observed of the whole population toward heterotrophic feeding in Burgos Bay. Probably this is a manifestation of the phytocenosis adaptation potential to the conditions forming now in coastal regions due to eutrophication.

5.7. NECESSARY CONDITIONS FOR PLANKTON BLOSSOMING

Burgas Bay - An Independent Center of Blossoming

The most visible after-effects of anthropogenic influence are the violated equilibrium in plankton communities and bentos organisms: "red tide", pests (deaths), extinction of bottom biocoenoses and so on. The live system violations are the final link in a chain of serious environmental changes of hydrophysical, hydrochemical and optical properties. As a rule, a considerable time is necessary for a live system to react to the transformation of quantity into new quality. This takes place because of the great adaptation possibilities of different types as well as of communities as a whole.

Violations of Black Sea ecosystem are obviously observed in the north-west part of shelf and seacoast. The ecosystem in open sea, limited by the circular Black Sea stream, for the present moment has no simply provable symptoms for anthropogenic influence, with the exception of some diseases in dolphins.

The most strong anthropogenic influence is registered along the west seacoast of Black Sea in the region from Dniester-Bug firth to Burgos Bay inclusive. The index is the "red tide" regularly observed since 1986 in these regions. The "red tides" worsen the recreation values of coast, cause deaths of fish, lead to a large quantity of organisms and changes in water optical properties, to intensified sliming of bottom and create conditions for hydrogen sulphide contamination.

Red tides in the Danube outfall have been observed since 1967 and repeated in 1969, 1974-1976, 1979-1982; near Varna from 1954 and then in 1958, 1963, 1974, 1975, 1984; in the north-west shelf from 1973. The spots of "blossoming" have different dimensions and configuration from 10 m to several km. The enumerated cases of "red tides" were provoked by development of one or two types of peridine weeds: more often Exuviaella cordata, rarer Goniaulax polyedra, G.polygramm and Prorocentrum micans, and also autotrophical feeding infusoria Mesodinium rubrum. A substance of toxic properties is released only by G.polyedra among the enumerated types. Experimental investigations showed that E.cordata does not release toxic substances and we have no data about other types.

The "blossomings" of E.cordata are known in many parts of the west seacoast and in north-west shelf. Blossomings of G.polyedra and G.polygramm have been observed many times opposite to the Rumanian coast. In 1979 in the center of Varna Bay (by materials of Vishnevski, S.L.) the biomass of these types was 105 g/cub.m and the number 4 x 10 type/l. There was a joint blossoming of P.micans with E.cordata in Varna Bay in November 1984. M.rubrum blossoming was found in the form of big spots in 1984 at Cape Kaliakra.

There was a "red tide" caused for the first time by the development of E.cordata in the Burgos Bay in May-June 1986. The center of blossoming was situated in Burgos Bay bottom, where its number in the upper one-meter layer reached 10 billions type/l, biomass 1 kg/cub.m, the chlorophyll content reached 930 mg/cub.m during culmination (3-4 June). Water was orange red with strong unpleasant smell.

The collected data of E.cordata "blossomings" in Black Sea allow to define several conditions necessary for the appearance of a "red tide": temperature not lower than 20 C, low saltiness (from 11 to 15 parts per thousand), a set of bionic elements, light more than 600 000 lux and absence of wave disarrangement (not higher than 3 bails rough sea). The last condition is especially important in the period of exponential growth at numbers from 10 to 100 million type/l. Low saltiness as an obligatory condition for a red tide defines its centers: Dniester- Bug firth, Shagani firth, The Danube mouth, Varna region (with the influx from Beloslavsko lake) and Burgos Bay. Burgos Bay internal part refreshes from the influx of two fresh water lakes (about 250 million cub.m per year). The spots and trains of "blossoming" at a considerable distance from coast are a secondary phenomenon connected with the wind transfer of surface water or concentration of cells at the boundary of water masses with different density (zones of contact between salty and fresh water or waters of different temperature). The existence of "blossoming" centers denies the dominating conception for the decisive influence of the Danube on "red tides" and on "blossoming" distribution by the Danube waters. Only the fresh water flow is comparatively accessible for control of all conditions necessary for a "red tide" appearance. By regulation, the system may be protected from a "red tide" appearance.

The "red tide" in Burgos Bay continues about two weeks, but cases of longer "blossomings" are known. The metabolites of E.cordata have not been a brake for the simultaneous development of accompanying types; diatomic Rhizosolenia fragilissima and peridines Prorocentrum micans, fluctuations that were negligible during "blossoming". At that, the first type in the initial and final phase of "blossoming", as well as in the open part of Bay, was 80% of the total phytoplankton biomass.

The fast decrease of blossoming was connected with pollution of E.cordata population with the ecto-parasitic flagellata. Thus, the formed suspended organic substance caused a sharp (ten times) increase of bacterioplankton and zooflagellates. In the population of E. cordata itself, a part of the unpolluted cells switched over the photoorganotrophic way of feeding on account of the dissolved organic substance utilisation (free amino acids and glucose). In this way, we may speak about two types of reactions of community to the anomalous phenomenon "red tide": on the one hand - a phenomenon of system self-cleaning by means of the parasitic flagellata, and on the other hand - a functional rearrangement of the mass type population switching over an other way of energy supply. The self-cleaning phenomenon may be interpreted as one of the way for biological cleaning of waters.

5.8. BACTERIOPLANKTON IN BURGAS BAY AND IN OPEN COASTAL WATERS

Bacterioplankton is one of the main components of sea ecosystems. It defines to a great extent the ways of organic substance transformations by its functional activity. The bacterial population along the Bulgarian seacoast of Black Sea is poorly investigated.

The data given below are from investigations during the 9th and 11th voyage of NIK "Academician L. Orbeli" (1981, 1982), the 6th voyage of NIK

"Vityaz" (1984), the 7th and 8th voyages of NIK "Rift" (1985, 1986). The number of bacteria was determined by the method of direct counting on epi-fluorescence microscope.

Bacteria production is determined by the radiocarbonic method and by two modifications of the direct method (by the use of antibiotics in check glass vessels) and by counting the time of micro-organism reproduction.

The number of plankton bacteria in Burgos Bay is unusually big. The minimal values are close to 1 million cells in 1 ml and the maximal exceed 10 million cells /ml. The biggest concentration of micro-organisms is characteristic for the surface layer. The number of bacteria smoothly decreases with the depth increase and at bottom layer it is ½ - 1/3 from their number in surface layer. Most high values are characteristic for the most polluted inner part of Bay where the water exchange is small. A gradual decrease of bacteria concentration is observed from coast to sea in the open part of Bay.

The mean volume size of bacterial cells changes in comparatively narrow limits from 0.10 to 0.30 cub. mkm, for autumn period - 0.183 cub. mkm on the average. With the increasing of depth the mean volume of bacteria decreases 1.5 - 2 times.

The spreading of bacteria biomass in the Burgos Bay reflects generally - their spreading in number. The biomass changes from tens and hundreds milligrams in a cub.meter to several grams at different horizons and stations. As a rule, it is typical for the biomass quantity to be from 0.5 to 1 g/cub.m in late spring and early summer. The total bacterial mass in surface waters as well as their number smoothly decrease and at the bottom is ½- 1/3 from the mass and number of bacteria in the surface layer. The biggest biomass quantity is inside the Bay and less in waters of its open part.

The abundance of bacterioplankton determines the high productivity in Burgos Bay. The rates of bacteria reproduction and micro-organisms production strongly depend on the momentum hydrochemical and hydrobiological conditions. Characteristic for the Bay inside are the values of bacterial production from 452 to 1308 mg/ cub.m in different layers or 902 mg/cub.m average, without vividly expressed phytoplankton blossoming late in spring. The bacterial production at surface is 2.7 g/cub.m at E.cordata "blossoming". In the central part of Bay the day production of micro-organisms is low. It is 360- 902 mg/cub.m average, changing from station to station and at different horizons from several tens of milligrams to values close to 1 g/cub.m. In the sea waters of Burgos Bay the magnitude of a day's production of bacteria varies in wider limits reaching in surface layers to 1 – 3.5 g/cub.m, as the average is 660 - 1700 mg/cub.m for the water column.

The vertical distribution of bacteria production corresponds to their concentration distribution. The determination of bacterioplankton reproduction rates shows that the bacteria generation time at temperature 17-24.5 C is from 6.7 to 30.1 h. At comparatively low temperatures from 8.5 to 10.5 C, the generation time increases. The destruction intensity of organic substances from bacteria is from 69 to 340 mg C/cub. m per day in late spring and early summer at different stations.

The ratio between primary production and bacterial destruction is under unit, i.e. the newly formed organic substance in photosynthesis does not cover the energy expenses even of bacteria. An exception is the period of rapid development of E.cordata when the magnitude of primary production exceeds the magnitude of bacterial destruction 30 times.

Open waters of Bulgarian seacoast characterize with the big content of bacteria although their concentration is several times lower than in the polluted regions of Burgos Bay. In general, moving away from coast and increasing the depth, the mean concentration of micro-organisms smoothly decreases. At a depth of 20 m (in the zone of intensive navigation), where permanent mixing of waters takes place, the mean number of bacteria for the water column is a million cell/ml, and the biomass- 282 mg/cub.m. At a depth of 30-40 m, the average number of the whole layer is smaller than in shallow waters, although in surface waters it may reach to 1.0-3.5 million cells/ml. The concentration of particles with dimensions of the order of 100 mkm, uninhabited with bacteria, is high in water. The role of the aggregated bacterioplankton and detrite, inhabited with micro-organisms, is inessential (22 thousand cells/ml or 2.7 kg/cub. m) in late spring. The biomass of micro-organisms in layers of largest abundance reaches 1000 mg/cub.m, and from 200 to 300 mg/cub.m average in the water column.

The number of bacteria in the richest surface waters is from 300 to 400 thousand cells/ml in late spring and the biomass is about 100 mg/cub. m in the most distant regions from coast at a depth of about 100 m. There the mean concentration of bacteria in the water column is 4-5 times lower than in the stations at a depth of 20 m (from 197 to 265 thousand cells/ml or 55-74 mg/cub. M).

The mean volume of bacteria in open coastal waters is from 0.113 to 0.264 cub. mkm and 0.178 cub. mkm in the water column.

The mean production of bacteria in the water column smoothly decreases with moving away from coast and increasing the depth in the end of May. The production magnitude of micro-organisms is 152-568 mg/cub.m at a depth of 10 - 80 m. At separate horizons the bacterial production may change considerably from several tens mg/cub. m to values close to 1000 mg/cub. m. The biggest values of production are typical for layers of maximum abundance of micro-organisms. These layers, like in Burgos Bay, are situated in the surface.

The intensity of organic substance destruction by bacteria for the water column in different stations is from 30 to 114 mg C/cub.m average per day (at values 15-179 mg C/cub. m, respectively for bottom layers and surface).

The ratio between primary production and bacterial destruction in open coastal waters, like in Burgos Bay, are under unit, i.e. the daily loss of energy exceeds its formation and entrance from phytoplankton. The situation is just the opposite only during "blossoming" of E.cordata. Then the primary production exceeds the bacterial destruction 5 times which speaks for the considerable influence of "blossoming" in the Bay on the communities in adjoining waters.

Open coastal waters of the Bulgarian seacoast characterize with a scarce amount of bacteria in autumn. The number of micro-organisms is at least twice lower than in the Burgos Bay at the same time and is a mean value for the water column at different stations 227- 570 thousand cell/ml. The fluctuation limits of the bacterioplankton number is between 150 and 700 thousand cell/ml. Smallest values are characteristic for the most distant points from coast. Comparatively high is the number of micro-organisms in the zones of soil masses discharge. The average volume of bacterial cells in open waters does not differ from this in Burgos Bay and the mean volume in all stations is 0.119 cub. mkm (at minimal and maximal values 0.082 and 0.179 cub. mkm) in autumn. This volume of bacteria is 1.5 times less than the characteristic dimension of cells in late spring in bay and in open coastal waters. Big changes are not observed in the volume of bacteria at different depths.

The biomass of bacterioplankton in open coastal waters is not big in autumn - from 50 to 100 mg/cub.m and does not exceed 50 mg/cub.m at the most distant stations. The value of daily bacteria production is biggest in the most inner part of Burgos Bay (332- 367 mg/cub. m) and in the region of Cape Atiya where it is 197- 414 mg/cub.m. Lowest values are typical for open coastal waters. The specific daily bacteria production in autumn does not differ from the values typical for spring and is 0.48-2.12. The generation time of bacteria at surface water temperature about 20 °C is 13-14 h.

The destruction intensity of organic substance from bacteria at the surface is 66-73 mg C/cub.m per day inside the Burgos Bay and 5- 6 mg C/cub.m in open coastal waters.

The primary production at the surface is 11- 484 mg C/cub.m in Burgos Bay, 11- 47 mg C/cub. m in open coastal waters (data of V.I. Vedernikov).

The development of pelagian microflora in Burgos Bay is average 5-10 times higher than in open sea. Correspondingly, the bacterial production is also more. The abundance of bacterioplankton in spring is considerably bigger than in autumn. The number and the mean dimension of bacteria decreases in autumn. The concentration of micro-organisms is biggest in the short period of intensive dying off of a big amount of phytoplankton (after "blossoming" maxima). Then the mean dimension of bacterial cells increases, the influence of bacteria on detrite sharply increases.

The ratio of primary production to bacterial destruction is less than unity during late spring and early summer when phytoplankton "blossoming" is not vividly pronounced. In the time of intensive development of weeds, the magnitude of primary production exceeds the magnitude of bacterial destruction from 5 to 30 times. In autumn the values of primary production are 2-9 times higher than the values of bacterial destruction according to estimations of samples from the surface layer.

5.9. MACROBENTOS

Bentos micro-organisms exert an essential influence on the formation of structure and ecological peculiarities at bottom, which are connected in a significant extent with the level of anthropogenic pollution in coastal regions.

Four basic zones may be differentiated in the basin by their contribution to macrobentos in the structure formation and the transformation of depositions. Big values of macrobentos biomass are established in the whole basin of Varna Bay from Cape Kaliakra in the north to Cape Emine in the south. The maximal depths in this zone reach 60-70 m. The largest width of the zone is opposite to Varna. Fields not great in area with increased macrobentos biomass are situated in north-east from Burgos and behind the isobar at 50 m opposite to Tzarevo.

The next zone by the extent of macrobentos development lies mainly between isobars at 50 and 100 m. Deep in depth (at the boundary with hydrogen sulphide), a poor zone is situated where the quantity of macrobentos organisms does not exceed 1 per sq.m.

The picture complicates under the influence of anthropogenic factors in the Burgos Bay basin. Two fields of comparatively small areas and low values of macrobentos biomass are observed in this region situated as follows: the one in proximity to Burgos and the other - north-east to Pomorie.

5.10. CONCLUSIONS

The analysis of data from observations done till May 1990, historical data and literature sources, expert estimations on the present state, and in some cases on the future ecological state of Burgos Bay, allow to characterize its state as unsuccessful.

The reasons are:

1. 133.2 thousand cub.m polluted and partly unpolluted wastewaters infuse in Burgos Bay per day as 92.12 thousand cub. m of them are industrial wastewaters. Additionally, the surface influx is 14.31 thousand cub.m water per day. About 13.45 tons oil products, 3.12 tons ammonia, 19.8 tons suspended substances, 22.6 tons organic substances per day enter the Bay.
2. Wastewaters discharged in Burgos Bay have a high redox-toxicity. To reduce them to the permissible standard, it is necessary to dilute waters at least 15 times. A toxicity may bring to a sharp increase

of pathogenic microflora, to an explosion of outbreaks and to "extinction" phenomena in bay in summer.

3. Pollution with oil hydrocarbons (OHC) places Burgos Bay in the category of extraordinarily polluted objects. The mean concentration in the whole bay exceeds tens of times the PLC. Background concentrations of OHC in open sea do not exceed 2 - 4 PLC.

4. The considerable permanent pollution with OHC and organic substance from drainages brought to dissolved oxygen deficiency in depth in the spring of 1989, at a comparatively low temperature and absence of a density jump, which reached to 15-20 per cents from the full saturation. This allows to predict that oxygen deficiency will increase in summer, when temperature increases,. This will cause deep hypoxia and "extinction" phenomena.

5. The hydrochemical regime of Bay strongly changes in the period from winter to spring, particularly the regime of biogenic substances toward their concentration increase to 1.5 - 2 times. This speaks for the serious hypertrophy of Burgos Bay.

6. The pollution of Burgos Bay with heavy metals (mercury, lead, copper) ranks it among the "strongly polluted". The concentrations of these metals exceed the background in open sea 10 - 20 times.

7. Primary production magnitude (from 1 to 2 g/cub.m) and plankton biomass (to 20 - 30 g/cub. m) in Burgos Bay bottom and Varna Bay region are typical for strongly eutrophicated waters. The primary production reaches to 30 g/cub. M and biomass to 1500 g/cub.m in the extreme cases of "red tide" in Burgos Bay.

8. The abundance of organisms after "blossoming" causes strong bacterial and heterotrophic plankton development. The ecosystem structure changes: weeds appear capable to photosynthesize and consume suspended organisms as well.

9. For the development of "red tide", a series of conditions are necessary: temperature not less than 20 C, saltiness from 1.10 to 1.59 p.p.m., abundance of bionic substances, light not less than 600 000 lux and rough sea less than 3 bails. Low saltiness defines the appearance of "red tides" in Dniester-Bug firth, Shagani firth, the Danube mouth, Varna Bay and the inner part of Burgos Bay.

10. The sharp decrease of traditional industrial fishing is connected with chemical pollution of water, "extinction" phenomena and use of bottom trails for industrial fishing.
11. Pollution with pesticides of Burgos Bay waters exceeded about fifty times the permissible limit concentrations in May 1990.
12. Pollution with phenol and its derivatives of Burgos Bay waters varies in different points from 0.02 to 0.08 mg/l and is average 0.054 mg/l, which exceeds the value of PLC from tens to hundred times.
13. The self-cleaning capacity of Burgos Bay for oil products is 0.7 ton-/per day in winter and to 2 ton/per day in summer.
14. The self-cleaning capacity of Burgos Bay, rudely speaking, for bionic substances, is about 50 ton/per day, as in winter is about three times less.

As unsolved scientific tasks have remained: the calculation of water and salt balance of Bay, determination of the velocity change in waters with regard to water balance, determination of the polluting substances mass balance, specifying the self-cleaning potential by components, study of the accumulative role of bottom with regard to pollution, bottom deposition role in the secondary pollution of Bay after rough sea, modelling of sliming at bottom, modelling of Mnemiopsis development, modelling and preparing of recommendations for recovery and naturally conformed exploitation of Pomoriisko lake, including salt production in the urban part and, of course, modelling of different scenarios for recovery of Bay ecosystem after several years.

5.11. RECOMMENDATIONS FOR MANAGEMENT OF THE ECOLOGICAL STATE OF BAY

Proposals for Immediate Implementation (May 1989)

1. To reduce discharges in the following places:
Pomorie - to put into exploitation the purification station by including a biogas stage and cascade of biological lakes;

Burgas

   a. To stop discharge of wastes from the stock-breeding complex into the collector of residential complex "Izgrev",

   b. To gather urban and wastewaters from "Fish Harbour" region in a collector and deviate them in town's purification station;

   c. To arrange an additional purification of wastewaters from "Neftohim" to the level of PLC, but not by diluting;

   d. The Navy - Atiya and Sozopol - to organize coastal equipment for collecting farming and urban wastewaters;

   e. Burgas harbour and Oil harbour - to organize coastal equipment for collecting waste ship's waters.

2. To study the possibilities and take measures for regulation and suspension of discharges from lakes, rivers and purification equipments if necessary in summer, to avoid blossoming and "extinction" phenomena.

3. To organize operational monitoring.

Proposals for Implementation in the Course of 1991-1992

1. Suspension of discharges exceeding PLC.

2. Organization of complex monitoring and its scientific software:

   a. To put into operation the first part of the Sea laboratory in Pomorie.

   b. To equip the Laboratory with necessary natatorial means.

   c. To equip the Laboratory with measuring and computing systems, means for reception, collection and transmission of information (including satellite) on the state of ecosystems.

   d. To ensure a staff in conformity with the Project.

3. To guarantee exploitation of the new technique of bulk weights unloading in the harbour of Burgas

4. To study the possibilities and utilisation of the flotation factory in "Vromos" bay, aiming at halt of harmful influence on the hydrochemical characteristics (high values of pH and radiation background).

5. To elaborate and realize preparation, installation and exploitation of closed systems for water using (CSW) in motor transportation enterprises in Burgos, Pomorie, Nesebar and Sunny Beach (in exploitation).

6. To discontinue the activity of small galvanic workshops operating without purification equipment (e.g. in Kableshkovo).
7. To introduce a system for collection, washing, packaging, storage and loading of metal chip, excluding the environmental pollution with lubricating-cooling liquids, oil products in machine building enterprises (in exploitation).
8. To avoid switch off of the biological stage in purification stations, inhabited places should be mapped for the content of oil products, heavy metals, chlorine organic compounds, surface active substances, etc.
9. To exclude chlorination of drinking waters by substituting with ozonization, UV treatment, etc.
10. To introduce a new technology in stock-breeding complexes in Aitos, Burgos, Pomorie and Kableshkovo by accompanying production of biogas fertilisers and production of milk products (elaboration of some technology elements).
11. To turn to naturally conformed farming with biological protection for some limited catchment area regions, including the decrease of large areas of mono-cultures by planting suitable trees.
12. To start the conversion of the machine building plant in Pomorie into a plant for purification equipments.
13. To introduce a severe control on the use of chemicals in agriculture mainly by financial policy.
14. To start accelerated planting of catchment area regions with broad-leaved types.

We should realize that the implementation of these proposals may stabilize only partly and for a short term the situation in Burgos Bay. That is why it is expedient, along with specifying the diagnosis on the state of Bay, to elaborate a long-term program to avoid worsening and for optimization of the ecological conditions in Bay.

A complex of scientific, technological, socio-economic and management tasks should be solved:

- switch over the purification equipments to anaerobic principle of work with accompanying production of biogas,
- switch over the public transport to electrical or gaseous,

- study the role of cultivated mussel and other sea cultures as filter,
- build and study the role of artificial reefs,
- study the possibilities for biogas production from biomass of sea cultures, development of sea cultures,
- study the exchange atmosphere-sea,
- decrease of discharges in atmosphere,
- division of Bay into zones with different purposes,
- conduct fundamental investigations allowing prognoses for the behaviour of ecosystems in real time,
- management of the regime and content of discharges in sea in dependence of the hydrometeorologic conditions,
- calculation and specification of the self-cleaning potential for management of the ecological situation,
- socio-economic and management undertakings allowing use on the principle: **one sea - one master,**
- active participation in the Association of Black Sea Towns and Regions "Clean and Peaceful Black Sea".

6

REGULATION OF SEA ANTHROPOGENIC BURDEN

The protection of Black Sea against unfavourable anthropogenic influences is a complex scientific, technological and management task. They are tangled into an exceptionally complicated knot:

•study of ecosystem changes under the influence of different anthropogenic impacts;

•methods of idealization and mathematical modeling, selection of parameters adequately describing the state;

•methods for taking decisions and their comparison with data from measurements in real time;

•creation of complex monitoring including systems for management and control of the behaviour of Black Sea ecosystems, processing and analysis of data, prognoses, elaboration of recommendations and control of their implementation and efficiency;

•standardizing the most harmful influences on the ecosystem;

•creation of a joint military ecological protection of Black Sea;

•introducing into practice cologically pure technologies in industry, agriculture and transportation;

•organization and start work on achieving the goals of the Association of Black Sea Towns and Regions "Clean and Peaceful Black Sea".

Science has available methods allowing to measure pollution level, estimate numerically the ecosystem burden by the separate polluting substances (CS), model the distribution, redistribution and transformation of CS at mass transfer and processes of water self-cleaning.

The state of the Program product created in 1989 with data bank "Black Sea", including data for water pollution sources and modern methods for regulating the quality of waters, allows formulation of some recommendations for preserving the Black Sea ecosystem against the most unfavourable anthropogenic influences. The different pollutants can be grouped by similar symptoms, in particular, by the character of their influence on the ecosystems and their reactions. We shall outline four most important pollutions:

1) Pollution of water with bionic substances: compounds of nitrogen and phosphorus, easily assimilated organic substances.
2) Pollution with suspended particles.
3) Pollution with strong toxic substances: (heavy metals, radionuclides, chlorine organic compounds, oil products, surface-active substances, dyes).
4) Pollution with redox-active substances.

The polluting substances of these groups have transformations close in their chemical character and equal influences in many cases in sea water.

6.1. POLLUTION WITH BIOGENIC SUBSTANCES

Basic sources of sea polluting bionic substances are:

- wastewaters from towns, villages and sea resorts;
- surface washing of soil from fields, meadows and gardens;
- unclean wastewaters from stock-breeding farms;
- beaches.

The bionic substances participate in the inner basin rotation of substances. In nature the exchange of substances is so balanced that in the system "aquatic surroundings-biota" a phase equilibrium exists by biogenic substances (E).

The exchange of bionic elements between biota and aquatic surroundings occurs mainly in the lowest trophic level, first of all with participation of micro-weeds which give the primary organic substance production. This organic substance is consumed by heterotrophic bacteria having a symbiotic relation with micro-weeds.

Mineral Forms of Nitrogen and Phosphorus

The anthropogenic influence stimulates the phytoplankton production, increases the velocity of organic substance formation and bacterioplankton production (depends on temperature), increases the trophy of water and so on. The negative after-effects of eutrophication are connected with "blossoming" of weeds and accompanying secondary effects. Part of these effects, connected with a change in the chemical and biological content of water, display immediately during "blossoming", other effects have a long-term character and lead to sliming of bottom followed by appearance of hydrogen sulphide zones, to growing of higher water vegetation, to a change in the content type, including the bentos organisms.

The "blossoming" of pyridine weeds is particularly dangerous for the sea surroundings (analogue to the green-blue weeds from fresh aquatic basin), producing algal toxins and forming quasi-reductive conditions in aquatic surroundings. Such conditions are favourable for reproduction of pathogenic and disease causing micro-organisms - carriers of different infections.

The "blossoming" of pyridine weeds takes place at a definite ratio of nitrogen and phosphorus mineral forms, and also at unbalance of oxidizing-reductive processes inside the basin. The cause is that green-blue and pyridine weeds photosynthesis is suppressed by hydrogen peroxide usually present in natural aquatic surroundings in concentrations 10 - 10 M. The other limiting parameters are temperature, saltiness, light and rough sea.

Obviously, to avoid "blossoming" of weeds, it is necessary to reduce abruptly the quantity of nitrogen and phosphorus bionic forms in the aquatic surroundings and also to suppress the falling of redox-active substances into water, since they violate the balance of oxidizing-reductive processes inside the basin. For the present moment, the only possible way for control of blossoming in Bay is by retaining the inflow.

6.2. EASILY ASSIMILATED ORGANIC SUBSTANCES

When the influx of easily assimilated organic substances exceeds over primary production, water bacterial pollution raises and the quantity of bionic elements increases with all accompanying effects of eutrophication. The influence of organic substances on the type of phytoplankton content may differ from the influence of pollution with nitrogen and phosphorus compounds. In consequence of water pollution with easily assimilated organic substances the dissolved oxygen deficiency increases, especially in layers at - bottom, which sharply worsens the state of higher hydrobionts.

Obviously, to avoid the negative after-effects, the quantity of easily assimilated organisms thrown out in sea should not exceed the primary phytoplankton production. Then the anthropogenic factor influence on the rotation of bionic elements will not be prevailing.

What practical measures are necessary to decrease the influx of bionic elements in sea?

First of all, to put an end to discharges of unpurified wastewaters from residential and stock-breeding complexes into sea. To reduce sharply the uncontrolled use of mineral fertilisers in farming industry in the region catchment area. It is well known that 30- 40% of the imported nitrogen fertilisers and 3-5% of phosphates fall into near basin. Besides, nitrogen in the form of nitrate penetrates easily in underground waters.

Purification of wastewaters of everyday life is done first of all biotechnological by using active slime. This method (by rational use of active slime) allows to transform the easily assimilated organisms into active slime. However, the biological method alone is ecologically insufficient for purification of wastewaters. There are at least three factors that make necessary wastewaters purifying after the biological block.

First: urban wastewaters contain a great number of viruses and bacteria, including pathogenic. These viruses and bacteria do not annihilate in aero-tanks and without any additional measures to finish purification they are discharged in basin. The lifetime of bacteria and viruses in sea surroundings is usually not long, however, it may increase sharply in surroundings polluted with organic substances and reduced oxidizing capacity (in the absence of hydrogen peroxide in water).

Lately, chlorine and its derivatives were used for sterilization of waters after the biological purification. In spite of the simplicity and low price, the use of chlorine for treatment of wastewaters is dangerous. To be efficient, a great amount of chlorine should be used for sterilization. The residual chlorine interacts with amines in the content of aquatic surroundings and composes toxic chlorine amines or with humus substances composes chloroform and other toxic chlorine containing compounds. Free chlorine is also strongly toxic with regard to hydrobionts.

Second: the biologically purified wastewaters contain great amounts of bionic elements, in particular, nitrates. Respectively, the discharge even of purified wastewaters of everyday life helps eutrophication of the basin.

The finish purification of wastewaters from bionic elements may be achieved by means of a third stage of purification - biological lakes. The bionic elements can be assimilated by sea cultures "sandwiches" of cultivated micro-weeds, mussels and higher water vegetation under control in coastal water-Basin. The obtained green mass may be utilized in different, ecologically pure and economically profitable ways. At water discharge from the biological lakes into open basin, no micro-weeds should be permitted in it. For this reason, water discharges from bio-lakes must be carried through a filter from the bottom layer.

Third: wastewaters of everyday life and industry may contain redox-active substances of reductive properties that do not stay long in aero-tanks. Falling into the basin together with the "purified" wastewaters, these substances can unbalance the oxidizing-recovery processes in the basin. The normalization of redox- state of wastewaters is possible in the first biological lakes, but there are more efficient ways for redox- detoxication.

The combination of biological wastewaters purification followed by hard ultraviolet radiation (UV-irradiation) is preferable. The ultraviolet radiation is particularly expedient for the preliminary treatment of mixed wastewaters of everyday life and industry entering for biological purification.

The widely applied UV-irradiation for purification of wastewaters is embarrassed by the lack of industrial production of quartz lamps for under-water work. We elaborate different constructions for mass production based on mercury lamps of low pressure.

This method for purification of wastewaters by using under-water quartz irradiators passed successfully the production tests and showed high efficiency in the intensification of biological purification. The cause is that under the action of quartz radiation a series of processes run in waste as well as in natural waters, leading to sterilization, detoxification and partly water cleaning from polluting substances. Except the bactericidal action, UV-irradiation interacts with the light-sensitive components of water. In particular, this radiation destroys such toxicants as aromatic hydrocarbons and some chlorine organic compounds. As a result of the photochemical processes in water, strongly active electron-excited particles compose, singlet oxygen, hydrogen peroxide and free radicals, including OH radicals. Under the action of these radicals and the participation of metal ions of variable valence and molecular oxygen dissolved in water, radical-chain oxidizing processes are running and oxidizing of different polluting substances, including heavily oxidized ones and poorly dissolved in water. Hydrogen peroxide O-radicals formed at recombination oxidize the redox-active substances with reduction properties - substrates of peroxide reactions. This leads to redox-detoxification and, consequently, to water self-cleaning.

Mixed waters, treated with "hard" UV-irradiation before the aero-tank, are more actively purified by the active slime. At that, the functional state of slime improves, the time of contact with purified water is reduced, purification indexes improve, the purifying equipment capacity considerably increases and purified water parameters also improve.

If biological lakes for finish purification are impossible for some reasons, then a combined biological purification is recommended with quartz radiation (submerged UV-lamps in the water flow at aero-tank entrance and exits of the secondary repository).

Economy is achieved at simultaneous treatment with quartz lamps and hydrogen peroxide in proportion 3- 5 1 30% H_2O_2 of 1000 cub.m. Thus the wastewater detoxificates from redox-toxins with reductive properties and the mass development of green-blue (pyridine) weeds is avoided in basin after the purification station. It should be remembered that quart radiation and the use of hydrogen peroxide do not clean the wastewaters from mineral forms of bionic elements (nitrates and phosphates), which are basic stimulators of phytoplankton "blossoming".

The ultraviolet radiation, treatment with hydrogen peroxide, blocks for anaerobic decay with accompanying production of biogas mineral fertilisers, cascade of bio-lakes with production of vegetation mass and fish, are basic technologies that should be introduced into the closed purification stations for wastewaters of everyday life, stock-breeding and also in local purification equipment enterprises for food industry and wine-cognac-producing industry. In the last two cases the economical effect may be added to ecological as well because of closed water use.

6.3. SUSPENDED PARTICLES

The causes for the appearance of a disperse phase in sea water are:

- soil erosion (surface and river);
- discharges of unpurified and with unfinished purification wastewaters;
- disappearing water micro-organisms;
- formation of calcium carbonate.
- Falling into aquatic surroundings, suspended particles cause a complex of unfavourable after-effects:
- decrease of water transparency and accompanying suppression of weeds photosynthesis (decrease of the primary production);
- domination of weeds in the ecosystem able to photosynthesize at weak sun light intensity and oxygen deficit, which is a peculiarity of green-blue (peridine) weeds;
- sliming of bottom;
- secondary water pollution Based on desorption of absorbed with suspended particles and bottom depositions of hydrophobic organic CS.

It was found recently that suspended particles absorb microcolloidal metastable particles with manganese oxides in a mixed valent state - Mn(III, IV). These particles are strong oxidizers. Falling on the bronchia surface they cause immediate death of fishes. The possibility to be toxic for those feeding with plants and for the zoofilters is not excluded (study in prospect).

First of all the suspended particles fall into sea uncontrollably by river tributaries. To avoid hyper-oxidized states together with aquatic media protection from suspended particles, it is necessary to exclude the technogenic pollution with manganese. At the same time water should not be polluted with calcium ions. This is achieved by introducing new technologies and local purification stations for different productions.

6.4. POLLUTION WITH STRONG TOXIC SUBSTANCES

These are substances of technogenic origin entering the aquatic media with wastewaters of industrial enterprises, through atmosphere, with rain waters flowing from towns after rain, directly from water transportation. As the "spectrum" of strong toxic compounds is unusually wide, it is expedient to group them by close properties.

HEAVY METALS are the best studied group of toxicants as cadmium, mercury, chromium, copper and other metals with variable valences. Some of the metals accumulate in living organisms, others participate in redox-processes in basin.

To decrease the pollution of sea surroundings with heavy metals from industrial wastewaters, different methods are used as electrochemical deposition, ion exchange resins, reagent materials. The blue vitriol (copper oxide), used in grapes protection, may exert negative influence on bay and coastal regions when the technology is violated.

Ions of transition metals participate in the creation of different complex compounds of ligands and suspended particles dissolved in water. In fact, the depositing particles in aquatic surroundings are as a "belt conveyor" that conveys metal ions from the water column into bottom deposits. Contrary to water, there often are recovery conditions in bottom depositions and metal ions with variable valence turn to a recovered form. At that, mercury ions recover to metal mercury. Most of the metal ions (except Mn and Cd ions) compose strongly related complexes with ligand groups and stay long in bottom materials.

Manganese circulates in the system water-bottom depositions.

The transition Mn in MnO in sea surroundings becomes first of all biotic by bacterial oxidation. In Mn oxidization, the formation of metastable hyper-oxidized states of microcolloidal Mn oxides is possible in a mixed valent state.

Radionuclides, as a rule also heavy metals, fall into sea with rivers and wastewaters as well as after accidents. Together with gamma-emitting isotopes (for example strontium and its row), the contamination of surroundings with alpha-emitting isotopes (plutonium, etc.) should be also taken into account. The last are especially dangerous because the GM-counters of gamma-rays, used for control of radiation contamination, are not sensitive to alpha-rays. Plutonium isotopes concentrate in complexes from different ligands dissolved and floating on the surface.

One of the environmental contamination sources with radionuclides, fluor and arsenic as well, are the different phosphoric fertilisers. The reason is that these elements are often in the content of natural phosphates. The work with phosphate materials with great content of radionuclides, imported from abroad, and the use of such "enriched" fertilisers is criminal and is subject to immediate severe ban.

An other "usual" contamination source with heavy metals and radionuclides are the power stations on coal. In spite of the low concentration of metals and radionuclides in coal, the enormous amount of burnt fuel leads to great absolute quantities of these elements in environment. The only exit for the present moment is the substitution of fuel with natural gas. The perspective lies in alternative energy sources with recovering resources.

6.5. CHLORINE ORGANIC COMPOUNDS

Chlorine organic compounds have industrial as well as agricultural origin. Due to their poor solubility in water, their stability to biodecays and chemical transformation, these compounds spread far in great distances and pass into underground waters. They possess the biggest coefficients of bioconcentration. The natural investigations show that these substances redistribute in the system "water-suspended particles-bottom depositions-biota" almost in correspondence with distribution coefficients in the system "water-octane".

From the industrial chlorine organic compounds, most toxic are the polychlorinated biphenol, chlorine phenols, dioxins. These super ecotoxicants are formed, together with benzene(a)pyren, dibenzofuran and other strong toxic compounds, in burning of solid wastes, in particular of plastics. That is why the utilization of plastics and other solid wastes without burning is an urgent scientific and technological task.

The production and use of chlorine organic pesticides continues in farming. They are hexachlorocyclohexane (CCCH), lontrel, 2,4-D and others. As an immediate step for environmental protection, including the Black Sea ecosystem, should be the complete suspension of chlorine containing and other pesticides stable to transformation in farming industry in the catchment area of Black Sea basin.

The control on the use and distribution of pesticides which disintegrate in the environment should be reinforced. The products of their transformation might be much more toxic for hydrobionts, in particular for fishes, instead of vermin we are fighting against. Such are, for example, most phosphorus - organic pesticides.

6.6. OIL PRODUCTS

Oil products are the most spread pollution in sea surroundings. Depending on the fraction and molecule weight, oil products (OP) form a film on the surface or compose water emulsion and sink in the bottom interacting with biota in a complicated way. The negative influence of OP on the ecosystem is connected with violation of the gas exchange between water and atmosphere, with bioconcentration and the effect of secondary pollution. The last circumstance is connected with the fact that OP accumulate in other hydrophobic CS in the anti-pole phase (in emulsions or bottom depositions). The aromatic hydrocarbons, entering into OP content, are toxic and concentrate in biota in accordance with the distribution coefficient in the system "water-octane".

Excluding accidental discharges, the OP sources in sea are: washing the urban streets, transportation enterprises, petrol stations, water transport, harbours, wastewaters of some industrial enterprises. Peculiar pollutants in coastal regions are the small fishing boats and ships, sports and tourist cutters, yachts. They are not supplied with systems preventing the release of considerable quantities of fuel and oils in water. In resort zones, systems should be created for increasing the ecological responsibility, for control on the exploitation of small tonnage cutters and yachts. Petrol stations, transport enterprises and motor-car services should be equipped with modern technologies with closed oil and water use, to solve the tasks for rational transport service. One of the conditions is accepting the principle: one village - one master.

The conversion to gas and electric power in public transport is an essential element in the protection strategy for resort zones and coastal waters from pollution with OP.

The purification of industrial wastewaters from the most toxic constituents of OP - aromatic and poly-unsaturated hydrocarbons can be efficiently achieved by the use of hard UV-irradiation, ozonization or by catalytic oxidation with oxygen or hydrogen peroxide. This should take place in the local purification station supplied with a system for closed use of oils and lubricant cooling liquids.

6.7. SURFACE ACTIVE SUBSTANCES

Surface active substances fall into aquatic surroundings mainly by everyday wastewaters and also with wastewaters from industrial enterprises producing PAV. The danger of PAV for the water ecosystem consists in their adsorption from surface by micro-organisms cells from a lower trophic level, which influences the production-destruction characteristics of the ecosystem and the water self-cleaning capacity.

The abiotic destructive oxidation of PAV is embarrassed and occurs first of all under the action of free radicals. In this respect the UV-irradiation, accompanied by wastewaters treatment with hydrogen peroxide, is perspective.

6.8. DYES AND VARNISHES

Dyes and varnishes are dangerous for aquatic ecosystems, as many of them are poorly dissolved in water and, therefore, the distribution coefficient in the system "water-antelope phase" is big. By their behaviour in the environment, dyes and varnishes are analogous to pesticides. A peculiarity of many dyes and varnishes (called dispersive) is their difficult biodegradation, which makes the biological blocks of purification equipments inefficient. The most efficient purification for dyes and varnishes is achieved in the combination of electrochemical methods with UV-irradiation and hydrogen peroxide treatment. Unfortunately, these methods are not used for the present moment on large scales. Some of the so called dyes dissolve well in water and possess well expressed oxidizing-recovery properties.

6.9. POLLUTION WITH REDOX-ACTIVE SUBSTANCES

Redox-active substances in wastewaters may have oxidizing or recovery properties. Among oxidizers, along with the free active chlorine, different nitro-compounds, metals in a higher state of oxidization, hydroperoxides, etc., may be present in wastewaters. As a rule, these compounds are toxic for the aquatic ecosystems. That is why their neutralization is necessary prior to the block for biological purification in the purification equipments.

Not less dangerous is the pollution of sea surroundings with substances of recovery properties. These substances except for their anthropogenic origin might be as a result of the live activity of green-blue (pyridine) weeds.

The unbalance of redox-processes in basin leads to the appearance of quasi-recovery conditions in water and has a series of negative after-effects:

The hydrogen peroxide deficiency is unfavourable for the development of fish's larvae. In the absence of hydrogen peroxide and the presence of substance recoveries, the metal ions, in particular copper ions, recover and compose biologically inaccessible and catalytic inactive complex forms. Because of the appearance of copper deficiency in the initial period of hydrobiont development, apo-ferments are formed not containing copper ions, which strongly influences the following development of hydrobionts. The substance recoveries themselves, in particular algal toxins, exert toxic influence on many hydrobionts. At the appearance of recovery conditions, the surroundings capacity sharply decreases for self-cleaning by the catalytic, in particular peroxidase processes of CS oxidization. Finally, disease-causing microflora reproduces in the quasi-recovery state of water.

7

STRATEGY AND TACTICS

The ultimate goal of our activity is unconditionally clean Black Sea with clean coasts and clean emptying rivers. This means ecologically clean industrial enterprises, ecologically clean farming, ecological harmless transport, absence of uncontrolled potential sources of pollution and so on. It is necessary to strive for this goal, however. Much can be done now, immediately, before the cardinal rearrangement of industry and agriculture.

First of all, to identify and localize, to control the pollution sources, including wastewaters from small powerful local sources. To organize quickly their purification taking into account the ecological requirements: sterilization and detoxification by ozonization, quarz ultraviolet irradiation, hydrogen peroxide, introducing a phase for anaerobic decay with accompanying production of biogas, utilisation of bionic elements in biological lakes to stop discharging the most toxic CS and suspended particles in the environment.

To divide the sea into zones for recreation, sport, fishing and sea cultures, yield of ores and minerals and places for waste discharge. At that, it is necessary to take into account the bottom relief, bentos state, regime of streams and meteorological conditions by seasons, bay ventilation and conditions for water mixing with all other hydrophysical, hydrochemical and hydrobiological factors, defining the mass transfer, utilisation and CS accumulations.

It is necessary to create a hierarchy system for monitoring and management of Black Sea ecosystem with telescopic options for the small bays and recreation zones. The program means capacity should ensure an estimation in real time for the self-cleaning components and volumes of CS of Sea as a whole and according to the chosen hierarchy of the local regions.

For the whole catchment area of Black Sea region as well as for its local regions, a system for control and management in taking measures at different accidents: from accidents in atomic power stations and big chemical plants to catastrophes with toxic and other loads during their transportation by ships, cars, trains and planes.

Satellite and other no contact monitoring, together with preliminary established powerful communication and computer net supplied with the corresponding software for work in real time, is the first main part of such a system.

Simultaneously, groups for ecological aid emergency should be organized.

The elaboration of norms is necessary for the permissible loadings with CS in the zones of wastewaters and earth's mass discharges. The self-cleaning capacity and the time necessary for CS to reach the surface layers should be estimated in the region of deep water outfalls of wastes from the big towns (Yalta, Sochi, Istanbul, etc.).

7.1. ASSOCIATION OF BLACK SEA TOWNS AND REGIONS "CLEAN AND PEACEFUL BLACK SEA"

Some scientists call the pessimistic prediction for the inevitable ecological change of Black Sea "ecological hysteria". Others believe that the self-cleaning capacities of sea will develop in such a way that the ecosystems will be able to take in the increasing pollution without much pressure, as simultaneously the quality of water and air will remain acceptable for people.

We wish to KNOW, not to BELIEVE.

We already know that during the next decade the Black Sea ecosystem will turn into a new state. However, we do not know details. The uncertainties in forecasting the parameters behaviour in bay are especially large, as well as in modelling of different scenarios. The present estimation of the value of these investigations is about 60 million levs (roubles) and about 12 million dollars in the course of three - five years.

Who will finance these investigations? The analysis of the present economical and political situation, the relations among USSR, Turkey, Bulgaria and Rumania, unequivocally shows that the Governments (and still less the Academies of Sciences) are not in a state to accept, co-ordinate and bring such a program to an end.

To our opinion, the interested party should pay. And these are the Black Sea citizens and departments using the Black Sea resources. In short, an Association of Black Sea Towns and Regions "Clean and Peaceful Black Sea"

Financing the reports-diagnosis on the ecological states of local regions and introduction of scientific recommendations in practice is also a task of the corresponding municipality in the frames of agreement for ecological and economical policy of the Association.

The establishment of a free economic zone and zones for free private initiatives for efficient usage of the natural resources of Black Sea is a good, real and easily attainable goal for the Association. The decision of Odessa of August 1990 for introducing a free economical zone in the town is an example.

Facilitating the contacts among people and their organizations in all possible levels and fields is worthy for the Association.

The establishment of regular shipping and air transport between Black Sea towns, facilitating the international tourism, including yachting, creation of necessary financial, Bank and communication conditions for development of business, are necessary for turning Black Sea into a peaceful sea.

Elaboration and realization of an educational program, distributed in newspapers, broadcast by the radio and television, video-clubs and variety theatres, cinemas, schools, colleges and universities, is a necessary condition for clean Black Sea.

Only by the establishment of a joint military ecological protection of Black Sea and the air-space over it may guarantee the accepted international regulations for environmental protection against illegal and unlawful pollution's

Undoubtedly, the Association of Black Sea Towns will favor the quicker preparation, signing and promoting the Convention for Black Sea.

Some History

Since 1986 we began to launch this idea in newspapers of some Bulgarian and Soviet Black Sea towns.

Initiator of the Association of Black Sea Towns became the district councils of Varna and Burgos in the spring of 1989 who accepted the recommendations of the International scientific program team "Black Sea", taken in the Seminar "Pomorie 88". After approval of the idea by the Bulgarian Ministry of Foreign Affairs, invitations have been sent to all Black Sea districts, regions and provinces. The first meeting of towns has to be held the same year in September in Burgos. However the local elections in USSR in the spring of 1990 postponed this constituent meeting.

The mayor of Odessa Valentin Simonenko invited the participants in "Eco Black Sea 1990" - representatives of Black Sea towns, on initiative of the Social-Ecological Union, to a two-day meeting in Odessa for discussing the main documents of the Association: Declaration on the aims and tasks and Statute of the Association. On 30 September 1990 the Declaration for foundation of the Association "Clean and Peaceful Black Sea" was signed in Odessa. On initiative of a group of members of the Bulgarian National Assembly the meeting was attended by observers from the Bulgarian Presidency. On behalf of Burgos and Varna the Declaration was signed by the councillor on ecology in the Presidency.

On 13 October 1990, a meeting was held in Varna of the Soviet mayors who had signed the Odessa Declaration with the new mayor' staff of Varna, members of the National Assembly, the councillor on ecology of the Bulgarian President Mr. Simeon Bozhanov, scientists of the group, journalists. Basic ideas of the Association Draft Statute and next year program were discussed.

The president of the district executive council of Burgos, Mr. Nedelcho Pandev, the mayors of Burgos, Pomorie and Nesebar supported the program

The idea for the Association and its program have the approval by a large group of members of the National Assembly from all parties.

The main purposes of the program are:

1. To inform all Black Sea citizens and sea users about the state of Black Sea ecosystem and the ecological problems in bays and coastal shelf as well.

2. To distribute the scientific recommendations for mastering the ecological situation among population, local authorities, directors and employees of departments and enterprises in regions polluting the bays.

3. To obtain estimation on the ecological responsibility of different pollutants of Black Sea and Sea of Azov to the end of 1991 and elaborate recommendations on this basis (on whole sea and regions) for naturally conformable managing, economical and technological decisions.

4. To promote implementation of the recommendations in practice.

CONCLUSION

Dearest citizens of Black Sea villages and towns!

The purpose of this booklet was to make you ponder over the future of your sea and convince you to take care of it, so that your children might breathe the air, drink water and bathe the sea.

In our opinion your joint efforts through the Association "Clean and Peaceful Black Sea" are the only way for taking immediate measures.

Governments are far, but mayors are always among you.

If you have undertaken steps to equip the purification station, modernize it, introduce ecological clean technologies wherever possible, closed circle of water use, scientifically grounded use of chemical preparations, turn to gas or electrical transportation, plant with trees the near naked mountain hills, then our efforts have not been in vain!

The necessary means, dear Black Sea citizens, will not come from the governments. They will be rendered by local and world business interest in normal environment, first of all tourism and transport.

You can find more on the problem in the next booklet.

20 November 1990

Sofia, Pomorie, Odessa

SOME CONCEPTIONS ON ECOLOGY AND CARE OF ENVIRONMENT

The term "ecology" has been suggested by the German biologist E. Heckel since 1869. It comes from Greek *"oikos"* (house) and *"logos"* (reason). Initially ecology meant the sum of knowledge about interrelations between organisms and environment, individual organisms or groups of individuals.

Ecology studies interrelations between living organisms and the environment of their inhabitancy. As a science it is founded on the achievements and methods of biology, chemistry, physics, mathematics, geography, geology and so on.

Recently people became aware that almost all fields of knowledge combine by one common sign - their importance for ecology.

The tasks of ecology as science are variable. Among them we shall point out the investigations on: organization and functioning of the world of organisms, mastering the biological resources, development of scientific criteria on the estimation of environmental quality, preserving standardized parts of biosphere, possibilities for management of ecosystems for naturally conform optimization of processes and their parameters.

Essential for human economic activities are: making decisions on management, minimizing the possible damages of environment, prognoses of negative after-effects in realizing certain unavoidable activities based on modelling scientific diagnostics of factors, processes and phenomena threatening human health and the state of natural systems.

The conception "nature" means "the environment and its infinite variety". To nature refer not only living organisms (plants and animals) but still components of the surrounding as well, including the lithosphere layer (the rock) of earth's crust. The most important raw materials and energy resources are concentrated there. There is also the so called "second nature": the totality of objects (goods) and phenomena that have not existed in ready forms but have been created as a result of public production. Some of these phenomena (processes) may have unfavourable after-effects on environment.

By virtue of natural laws, all kinds of human activities lead to formation not only of necessary substances but to waste products as well. These substances are foreign for the environment where the living organisms are developing. For this reason they are called "xeno-biotic" ("*xenos*", Greek origin, means foreign). If in the biochemical cycle of a given living organism xeno-biotic substances participate, then the normal exchange processes are embarrassed or destroyed (as it happens in all functioning systems). This has negative after-effects on organism and population.

Annually hundreds of organism types disappear from the face of our planet, represented in numerous populations.

Man reshapes nature and its concrete landscapes at high rates, adjusting them to his needs. History shows that in spite how much unfavourable for nature is a given process, its functioning is almost never ceased if it is beneficial for mankind. This contradiction may be overcome if the processes in question are modified or performed with intensity unchanging the environmental parameters to unacceptable ones for the inhabiting organism types. This requires ecological knowledge and skill of society in order to combine technologies with the laws of nature.

Earth's biosphere consists of zones from lithosphere, hydrosphere and atmosphere that are inhabited by organisms. The lower boundary of biosphere in the continental section of stratosphere is at a depth of 2 - 3 km, and in the oceanic section at 1- 2 km, where is limited by the temperature increase with depth. The upper boundary of biosphere spreads to a height of 20-25 km and is limited by the known "ozone layer", which protects the living cells from ultraviolet rays.

The live substance is concentrated mainly in earth's hydrosphere and on the surface of lithosphere. About 80% of still substance in the upper part of lithosphere is processed or treated by the organisms living there. To accomplish such grandiose activity, the organisms should have a significant mass. Indeed, the total mass of living organisms on earth is about 2400 billion tons and the part of them inhabiting land is 99.87%.

The biomass of organisms inhabiting oceans, seas and rivers is negligible compared to the mass of organisms inhabiting land. At the same time, however, the annual production of live substance of both sections is commensurable: in land - 180 billion tons, in ocean - 80 billion tons.

Life appeared on earth more than 3.5 billion years ago. A part of the production of live substance remains permanently in soil, inters peats, in depositions of aquatic basin and converts gradually in natural gas, oil, coal, bituminous schist, etc. The primary source of energy for almost all live creatures on earth is sun. The phenomenon "life" is a complicated process of reproduction, an incessant exchange and transformation of energy and substances with environment.

Environment characterizes with an enormous variety of parameters, many of them are factors of life, i.e. ecological factors. The latter divide in biotic factors (of live nature) and abiotic factors (of still nature). Organisms react to the ecological factors of the surroundings with specific adapting reactions.

Every living organism needs a definite combination of temperature, moisture and nutrition substances etc., in a suitable regime. This regime is determined by the "lower" and "upper" boundary values of these factors. The wider the intervals, at which the organism may exist and reproduce, the bigger its adaptive possibilities and its tolerance with respect to external conditions.

The abiotic factors of environment are numerous and a part of the characteristics of earth's surface, atmosphere, soil layer and aquatic surroundings: pressure, density, temperature, transparency, light, saltiness, chemical content, radioactivity and so on.

Biotic factors of the surroundings are the characteristics of organisms, populations and their relations.

Plants and weeds are one basic organic group, which creates the primary organic substance, food and necessary energy for development of vegetable organisms.

Animals show preference to the food content. There are types among them eating only one type of organisms (monophages). Others, called polyphages, consume a large scale of types.

Besides the mostly spread type of relations in the animal kingdom - the ratio "victim-beast of prey", there are relations of parasitism, commensalism, ammensalism (one of the types does not develop in the presence of the other), mutualism (the type develops only in the presence of the other), neutralism and so on.

In the biological community composed for many years in a given geographic area (biotop), all its members are mutually dependent and necessary. The disappearance of the natural enemy may lead to suppression and even to extinction of the type - victim. A similar example in the near past was the mass extermination of wolves in Canada, whereby the deer herds were practically destroyed by diseases that led to degeneration of the type.

Plants develop normally at optimal correlation of mineral components in soil. A suppressing influence may have not only their deficiency in soil layer but the exceedingly high concentrations of some of them as well. As the deficiency of these elements does not compensate the excesses of others, the surroundings concentration of substances is the limiting factor for the development of vegetation types.

Hence, every organism is adapted to live in definite environmental conditions. The change of its parameters beyond definite limits suppresses vital functions of the type and may lead to its death. The totality of environmental parameters, limiting the conditions for existence and reproduction of a given type, represents the so called "ecological niche". Every type occupies a definite place in the environment, which is conditioned by its necessities (food, inhabitancy, reproduction). The requirements of different organisms to the parameters value of their ecological niches can be differently strict. Some organisms may develop in narrow and others in wide variations of factors. The adaptation capacity is called ecological valence or flexibility of the type.

All live organisms exist in nature only in the form of population or multitude of individuals from one and the same type, inhabiting a definite space (biotop). The reproduction is distribution and exchange of genetic information.

Every type of population possesses a historically composed structure (age, sex, space). The populations have specific indications which characterize them as multitude. For example, the number of individuals in a population, total biomass, number of individuals in a unit volume and unit area (density), consumed energy, velocity of feeding, velocity of disappearance and so on.

The totality of jointly living different types of organisms in the surroundings in which they exist is called ecological system (ecosystem).

The continental aquatic ecosystems divide in two big groups in dependence of the surroundings dynamic characteristics: lentic or quiet (marsh, lake) and lotic or strait (brooks, rivers). The big sea and ocean basin may combine section types of different dynamics.

Several main zones may be separated in every lentic or sea basin: litoral or shallow water, where light penetrates to bottom; limnic, where still active light penetrates (to 1% of the initial flow); profound - no penetration of sun light. Batial, abyssal and hadal constituents of basin's deep water part release in seas. Water space over the litoral area is called neritic zone and over the deep water - pelagian zone.

Depending on the place of inhabitancy, aquatic organisms are called:

- bentos: the totality of bottom organisms that have a fixed (immobile) way of life or move on or close to the bottom;
- periphyton: animals and plants attached to stems of higher plants and weeds, raising over bottom;
- plankton: free organisms passively floating with the streams of aquatic surroundings (the dimensions of organisms are microscopic or not big);
- nekton: independently swimming, moving actively organisms;
- neuston: organisms living in surface microlayer (larvae, weeds, etc.).

The living organisms which compose the biocoenoses of aquatic basin assimilate in a different way their necessary nourishing substances or quantity of energy. Unlike plants, animal organisms cannot photosynthesize or chemosynthesize and for this reason they use the sun energy indirectly consuming photosynthesizing organisms, other organisms or organic matter.

A chain of types is composed in biocoenosis, passing substances and energy to each other - trophic or nourishing chain.

Plants and weeds are producers of the organic substance as they are autotrophic organisms - create the primary organic substance from inorganic substances in photosynthesis. All rest types (animal) are heterotrophic organisms, they live on the substance of other organisms and for this reason are called consummates (consumers). They are two categories: on vegetable feeding and beasts of prey. The latter are consumers but of second level, and there are beasts of prey from the third "stage" of the chain, living on smaller beasts of prey.

The trophic chains may be different in length. They may bind collaterally with other chains, creating trophic nets determining basic energy elements and substance rotation in the ecosystem.

One important group of organisms is called reducers. Its representatives decompose the still (the dead) organic substance to primary mineral components and elements: gases, water, salts. Bacteria are the most various and efficient reducers. They divide into aerobic and anaerobic. The first use free oxygen (atmospheric or dissolved in water) for breathing and the second take it from different compounds. Anaerobic bacteria play a very important role in aquatic basin by decomposing the group of sulphate anion.

Sunlight penetrates to a definite depth in the aquatic ecosystems. Photosynthesizing organisms reduce quickly with depth as well as photosynthetic production. Because of that, the breathing processes connected with oxidization and disintegration of the organic substance intensify. For that reason such a basin where plants increase their biomass, i.e. where photosynthesis dominates over breathing, is called euphotic. The production of photosynthesizing organisms depends on the presence of bionic elements in water.

The systems rich in bionic elements are called eutrophic. When the content of biogens in basin increases too much, the quantity of biomass may reach to exceedingly high values. Then the equilibrium of mass exchange processes breaks in basin and organisms that have not succeeded to mineralize themselves begin to pollute the aquatic surroundings.

In a well functioning aquatic basin, the transformation of dying organic substance is along the line of its oxidation, decay and mineralization. The mineral components may be used again in the process of photosynthesis. Thus the processes create a biochemical rotation of substances and energy. Two such rotations are distinguished: big or geological and small or biological.

The geological rotation continues hundreds, thousands and millions years. During this period a part of the dying organic substance, that inters (fossilises) in bottom depositions, transforms in organic minerals. The organic minerals are found in a concentrated form as natural gas, oil and coal or in the type of a dispersed phase among rock matrices (water dissolved gas, sorbed oil or micro-oil, kerogen). At volatilization of depositions and rocks, fallen on earth, the organic minerals also oxidize and mineralize. A minimal part of the organic substance dispersed in rocks redeposits in aquatic basin and then again includes in the rotation.

The small rotation, which is a part of the big one, is running in a single ecosystem. The bionic elements accumulate initially in plants, pass into organisms consummates, and after their disappearance to organisms reducers. The latter decompose the biomass to simple mineral forms, which return into hydrosphere, atmosphere and soil.

For some elements, for example carbon and oxygen, exists the so called biologic-technical rotation appearing after the intensive use of excavated fuels. The use of these fuels increases the quantity of carbon dioxide of technogenic origin in atmosphere and intensifies the greenhouse effect. The main part of carbon on earth contains in carbonate minerals, constituting limestone, dolomites and some other types of rocks. The other much smaller part spreads among different organic compounds, including in excavated fuels. A comparatively small quantity of carbon is concentrated in the biomass of plants (500 billion tons) and animals (about 5 billion tons). Its rotation is comparatively fast. Nitrogen is 4/5 of air. This gas does not burn and does not feed burning and breathing. For this reason is in such great quantity. But a part of it contains in the biomass of all types of organisms, in water and soil. After dying of organisms, bacteria decompose the tissues containing nitrogen (ammonification and nitrification). At that ammonia, nitrates and nitrites are composed. A definite part of nitrates and nitrites is assimilated by plants and the other one recovers to free nitrogen which passes into atmosphere. This is the small nitrogen rotation. The free nitrogen absorbs again through the roots of plants due to tuber bacteria. In the big nitrogen rotation its inorganic compounds are included, spread in land and atmosphere.

Phosphorus is one of the most important bionic elements. Its natural rotation also has a small and big cycle. But it has also a biologic-technical rotation due to the widely used washing preparations and phosphate fertilisers.

ANTHROPOGENIC INFLUENCE

Mankind processes or uses in different ways 55% of land territory, 13% of river waters, replaces up to 4000 km^3 soil and rock masses annually and extracts more than 100 billion tons of ore from earth's entrails, burns more than 7 billion tons fuel, releases over 500 million tons of chemical substances (1/3 of them go and enter in hydro- and atmosphere). About 500 thousand types of chemical compounds are used. About 40000 of them are harmful for human health and about 12000 are directly poisonous.

The influence that mankind exerts over biosphere leads to:

1. Changes in landscape structure of earth's surface create new morphological elements;
2. Changes in biosphere content, characteristic of matter rotations, mass exchange balance between the geosphere covers of planet;
3. Changes in the energy balance in separate regions and in large scales;
4. Changes in the content of biota type, including disappearance of some types.

Pollutants of the environment are wastewaters (from industrial productions and stock-breeding complexes), bionic substances, inorganic acids and salts, solid state waste products, radioactive substances and so on. Pollutants are also all physical fields of thermal, light, noise, electromagnetic and so on, action and intensity exceeding the natural values. Chemical pollutions are of the most harmful. Mutations of some bacteria and viruses may convert into a serious biological pollution.

Pollution is a multilateral process with "branched" after-effects. Its after-effects not always quickly display. Not infrequently, obvious consequences of pollution are preceded by a long "incubation period" of hidden pollution. Because of that, significant efforts are applied today for a timely direct or indirect diagnostics of pollution's The destruction of biosphere, loss of resources and degradation of the environment leads finally to worsening the physiological state of man as well as the moral and psychological health of society.

A special anxiety arouses the hydrosphere polluton. The volume of the World Ocean is only a little more than 0.1% from the volume of the planet.

The water layer width, uniformly distributed over the whole earth, constitutes only 0.03% of its diameter - one thin membrane on its surface! Water is the only natural liquid that appears in a considerable quantity not only in hydrosphere but also in lithosphere and atmosphere (as water vapour). Existing in three aggregate states, water has enormous importance for the running of significant processes, including the formation of relief and appearance of life on earth. Without water, a great part of the technological processes used by mankind annually to an amount of 3000 km^3 fresh water, is unthinkable. At that 150 km^3 water are irreversibly taken out of rotation every year. Most water is used in farming (the biggest share of water irreversibility falls again on it). For the raising of one ton wheat 1500 tons of water are necessary, for 1 ton cotton - 10000 tons of water. In industry, the processes of chemical synthesis in rubber production absorb most water (to 3000 tons of water for 1 ton production). Water is already a deficient product for many fields and geographical regions in the world and, in perspective, a global crisis is outlined with consequences for mankind and biosphere difficult to be predicted.

Basic pollutants of hydrosphere are the atmosphere by rainfalls and direct exchange, wastewaters of everyday life and industry, wastewaters of stock breeding complexes.

In general, sea and ocean basins are polluted mainly by atmospheric streams of mass transport and river flows. Only in the Rhine River about 940 tons of mercury, 1040 tons of arsenic, 1700 tons of lead, 1400 tons of copper and 13000 tons of zinc fall annually. The Danube brings about 3000 tons of nickel, 14000 tons of manganese 36000 tons of oil products together with many other heavy metals, pesticides and nutrients (ammonium, nitrates, nitrites and phosphates) annually in Black Sea. About 15 million tons of oil and oil products fall annually in different ways in the World Ocean. Every ton covers 12 sq.km of water surface with a thin membrane which embarrasses the mass exchange between Ocean and atmosphere and harms directly the life conditions of micro-organisms. About 1/5 of the World Ocean basin is permanently covered with oil and oil products.

The bionic elements - nutrients stimulate the mass reproduction of definite groups of plankton organisms in basin of all types (mainly weeds and grassy vegetation). Reproducing fast, these organisms exhaust the water dissolved oxygen after dying and it becomes deficient in the bottom layer of big basin areas, where an anaerobic decay begins with hydrogen sulphide release. This whole complex of processes and phenomena is indicated with the term "eutrophication" of basin.

For the present moment, mankind cannot stop industrial wastes discharge in the environment. However, their quantity begins to be limited at least in economically developed countries. There are already national and international organizations recommending a fixed upper limit of waste products presence, different in content, in different geosphere objects. Thus, for example, the Experts' Committee of the World Health Organization has a list of the permissible pollutions in yearly, periodic and daily intervals. Some countries use norms based on temporarily permissible concentrations (TPC). For the present moment, the basic index used for estimation of the surroundings quality is the excess of permissible limit concentration (PLC) of harmful substances.

There are two main ways for determination of PLC: experimental and theoretical. The first is used for fixing the estimation for PLC and the second - for TPC. There are also methods for an express estimation of PLC by deducing the dependencies between concentrations of admixtures and the time of their action based on short-term (monthly) experiments.

As the different substances may exert, in general, similar unfavourable influences on the environment and its inhabiting organisms, intensified by non-linear effects, a special term was included - "summing effect". According to arithmetical summing, the quality of surroundings is appreciated as normal if the sum of concentration ratios of polluting substance to its PLC is less than a unit. It should be noted that this estimation does not record the joint effect of pollutants sum that sometimes can be intensified many times.

The extent of permissible limit contamination of aquatic surroundings is determined not only by current parameters but by the ecosystem capacity to neutralize pollutants as well. It is determined by the biotic component which requires norms for estimation of the permissible limit ecological loading.

Timely registering is necessary for unwanted deviations in the surroundings quality and finding its optimal parameters. Nowadays, this surroundings whose parameters correspond to the parameters of the ecological niche of mankind, and do not contradict the scientific and technical progress of society, is often accepted as "qualitative". However, society is inhomogeneous and consists, in reality, of multitude of communities whose parameters are sometimes rather different. Correspondingly, the parameters of their ecological niches also differ. Moreover, it becomes obvious at a careful analysis that some of these ecological niches are incomparable. Some ethnic and racial human groupings represent rather strictly organized communities that exist well only in given surroundings. The parameters of these surroundings may not correspond at all to the indexes typical for the ecological niches of industrially developed communities. The examples with "primitive" tribal communities from polar tundra or tropical forests show that their evolution in conditions of "civilized surroundings" takes place with a loss of the community primary identity. In most cases this degrades the primary community. A similar adaptation to ecological niches with new parameters may have fatal consequences of irreversible character for the community and very often for the individuals that constitute it. Still more conflicting and incompatible situations appear in the combination of parts from ecological niches of thousand plant and animal types with the ecological niche of the technological human community.

It is hard to estimate the "optimal surroundings" in the presence of so many examples only on criteria useful for *"Homo Sapiens"* type. Such approach is extremely "Homo-centric", intolerable to the rest types of organisms and, in general, non-ecological.

These and a series of other problems in ecology still have not found a simple and good solution. In many cases this solution is delayed due to the lack of sufficient and useful information. The attempts to organize "ecological monitoring" face many difficulties of technical and moral character.

THE MOST ESSENTIAL STEPS

The impressive development of mathematics, physics, biology, chemistry, engineering sciences and technologies in the last ten years, the powerful computerizing, give the possibility to transfer ecology into a real science for management of ecosystems:

A. The preliminary wise idealization of the ecosystem, mathematical modelling of processes and the interactions in it, development of different populations and their mutual influence, rotation and exchange of substances and energy, are the first step;

B. The second step is the simultaneous search for analytical and numerical decisions and their comparison to experimental data;

C. Formulating and solving the reverse task for mathematical hypotheses and finding the numerical values of unknown coefficients and parameters with estimation of their reliability, are the third step;

D. The fourth step is the elaboration of a system for ecosystem management.

The system includes the following blocks:

1. Monitoring;
2. Block for solving equations from second and third steps;
3. Block for comparison of decisions with data from monitoring and their correction, working in real time;
4. Block for table, graphic and mapping (two- and three-dimensional) visualization of the so obtained solutions;
5. Block for playing the different variants of future behaviour of the ecosystem or some of its sub-systems depending on the choice of ruling parameters;
6. Block for management recommendations;
7. Archival block.

MANUSCRIPT OF INFORMATION BULLETIN No 1 OF THE ASSOCIATION OF BLACK SEA TOWNS AND REGIONS
"CLEAN AND PEACEFUL BLACK SEA"

No 1
NOVEMBER 1990
ODESSA, POMORIE
To
ALL MAYORS OF BLACK SEA TOWNS AND REGIONS
DEAR SIRS!

As you know "The Odessa Declaration" has been adopted on 30 of September 1990 with the objective to create the Association of the Black Sea Towns and Regions " Clean and Peaceful Black Sea".

We inform you that Burgos region and the towns of Pomorie (Bulgaria) and Odessa (USSR) have agreed on the membership fee at the rate of one lev (rouble) per citizen.

I have the pleasure to invite you to participate in the first General Meeting which will take place at the end of January 1991 in the town of Pomorie, Burgos region, Bulgaria, and we have the permission from the town's citizens.

We are looking forward to meeting you in the town of Pomorie!

Happy New Year!

Sincerely yours,

Valentin Simonenko

Mayor of Odessa

22 of December 1990
City of Odessa

Content

Odessa Declaration
Status Draft
Programme Draft For 1991
Community's Presidential Council Membership's Draft
Appendix:

1. CONTENT OF THE "REPORT- DIAGNOSIS"
2. PROPOSALS FOR BUSINESS, FUNDAMENTAL AND APPLIED RESEARCH
3. CLEAN AND PEACEFUL BLACK SEA

ODESSA   30 SEPTEMBER 1990
DECLARATION

for foundation of the Association of Black Sea Towns and Regions
"CLEAN AND PEACEFUL BLACK SEA"

The representatives of Black Sea towns come to the following conclusion in their first international meeting:

The efforts undertaken until now by the governments of the four Black Sea countries to avoid Black Sea pollution do not give noticeable results;

In reality many Black Sea regions are in ecological crisis, The Sea of Azov is dead;

The sea basin pollution continues, exceeding many times the sea capacity of self-cleaning;

The recreation resources of Black Sea Coasts are facing death;

The traditional fish supplies are almost completely lost;

The social, transport, trade, scientific, cultural relations among Black Sea towns develop very poorly.

Expressing their great anxiety about the fate of Black Sea, the representatives of Black Sea towns declare their intention to unite and oppose the coming global catastrophe of Black Sea with co-ordinated policy for control and mutual responsibility in order to save the unique creation of nature for man and mankind, a source of wealth for the peoples living here.

Declaring the foundation of the international Association "CLEAN AND PEACEFUL BLACK SEA", the towns make the following their aim and task:

Co-operation of the intellectual potential, material and finance means, authorities functions and management efforts for saving and recovery of Black Sea;

Developing the contacts among Black Sea towns, raising the ecological culture, elaboration and implementation into practice of joint policy for control of Black Sea region pollution, for management of the ecological situation;

Establishment of long-term social, cultural, scientific and trade relations.

The humane aim of the Association CLEAN AND PEACEFUL BLACK SEA will be firmly pursued by the Black Sea towns!

FROM THE TOWNS:

BATUMI	Sulico INASHVILI,
	First vice-president of SM of Adjaria
BURGAS	Simeon BOZHANOV,
VARNA	Councillor of the Prime-Minister
	of Bulgaria on ecology
MARIOPOL	Stanislav KOSHELEV,
	President of the Executive Com.
ODESSA	Valentin SIMONENKO,
	President of the Executive Com.
SUHUMI	Djansuh GUBAZ,
	President of the Executive Com.
KHERSON	Valentina SHTERBINA,
	President of the Executive Com.
YALTA	Svetlana LEVCHENKO,
	Vice-president of the Executive Com.
SEVASTOPOL	Arkadii SHESTAKOV
	President of the Executive Com.

Draft
Statute of the Association of Black Sea Towns and Regions "Clean And Peaceful Black Sea"

Chapter One: Name and Headquarters

Article 1. The Association of Black Sea Towns and Regions "CLEAN AND PEACEFUL BLACK SEA" is a non-governmental international organization with an ideal purpose.

Article 2. The Association is a juridical person with headquarters in the town of Pomorie, "Peyo Yavorov" str. No 1.

Article 3. The Association founds its centers in Constanza, Rumania, Odessa, Sevastopol, Gelendgiê, Suhumi, Batumi, S.I.S., Istanbul, Turkey, Varna, Burgos, Bulgaria.

Chapter Two: Aims and Means

Article 4. Aim of the Association is the protection of purity and peace in Black Sea region, realized by implementation into practice of naturally conformable managing, economical and technological decisions by integration of human, intellectual, material and other resources.

Article 5. The Association assists the governments of Black Sea countries in the elaboration and agreement on a Convention for defense and protection of Black Sea from pollution, and for joint management of the ecological situation.

Article 6. The Association co-operates with the governments of Black Sea countries and with international organizations whose aims are peace and protection of the environment, with all juridical and physical, local or foreign persons, who accept its aims.

Article 7. The towns create free economical zones or zones for free economic initiative.

Article 8. The towns forward the contacts among people, public and economical organizations on any level and in all fields.

Article 9. The towns create or help in the development of regular sea, air, motor and railway transport in the region.

Article 10. The towns together or separately create favourable conditions for development of international tourism, including yachting.

Article 11. The Association helps the insurance of necessary financial, customs, communication and other conditions for development of business in the region.

Article 12. The Association co-operates with the governments of Black Sea countries for the establishment of a joint service for ecological protection of Black Sea and the air-space over it, for security and control in performing the international rules, for environmental protection against illegal pollutions, for observance of the international agreements of naturally conformable use of sea resources.

Article 13. The Association participates in financing and executing the program for ecological training and education.

Article 14. In co-operation with the governments and towns the Association creates legislative, normative, financial and resourceful conditions for implementation into practice of the recommendations for management of the ecological situation all over the sea, as well as in separate regions.

Chapter Three: Forms of Activity

Article 15. To fulfil its tasks and achieve the aims, the Association:

- point 1.   Holds symposia, seminars, conferences, demonstrations, meetings, public discussions, inquiries of public opinion, TV-bridges, other scientific and social propaganda actions;
- point 2.   Finances publishing of information bulletins, informational, scientific, popular scientific, publicists, advertising and other printed, audio, cinema and video-production;
- point 3.   Delivers regularly its materials to mass media;
- point 4.   Organizes lectures, exhibitions, videos, film-shows, as well as scientific and research expeditions;
- point 5.   Performs   ecological,   ecology-economical   expertise, sociological and other researches;
- point 6.   Works out methodology for expert estimations of big projects for nature transforming;

- point 7.   Aids the legislative organs in the acceptance of normative acts in the field of ecology;
- point 8.   Organizes investigations on estimation of the ecological responsibility of subjects polluting environment and elaboration of naturally conformable management, economical, technological decisions;
- point 9.   Co-operates in the elaboration of new scientific ecological conceptions, finding and study of ecological problems for removal of prior-to-crisis and crisis ecological situations;
- point 10. Finances scientific, construction, project and implementation works;
- point 11. Creates scientific teams;
- point 12. Organizes competitions, establishes prizes, awards;

Chapter Four: Structure and Organs

Article 16. Organs of the Association:

1.   General Assembly.
2.   Board of managers.
3.   President.
4.   Co-ordinator.
5.   Executive director.

Chapter Five: Membership

Article 17. A member of the Association can be every town or field from the coasts of Black Sea and the Sea of Azov, who accept the Statute and pay the annual fee defined by the General Assembly. The members of the Association participate in its work through its representatives, authorized according to the rules of the local legislature and this Statute.

Article 18.

1. The members of the Association are accepted by the General Assembly or by the Board of Managers by written application.

2. The decision of the Board of Managers is taken by a majority of two thirds of all members with suffrage and is subject of approval by the General Assembly.

3. If the decision of the Board of Managers is rejected by the General Assembly, the membership discontinues from the date of the Assembly. A new application for membership may be given not earlier than six months.

Article 19. Every member of the Association has the right:

- point 1.   To participate in the General Assembly, to vote in taking decisions,
- point 2.   To express freely own position on questions put to discussion,
- point 3.   To choose and be chosen in the organs of the Association,
- point 4.   To use the property and activity of the Association according to this Statute, the decisions of the General Assembly and the Board of Managers, and local legislation,
- point 5.   To make proposals in the General Assembly and the Board of Managers.

Article 20. Every member of the Association is obliged:

- point 1.   To abide this Statute,
- point 2.   To pay regularly the membership fee,
- point 3.   To assist in the achievement of the aims and tasks of the Association,
- point 4.   To execute the decisions of the Association organs in accordance with this Statute, local legislation and local conditions and possibilities,
- point 5.   To look after the property of the Association.

Article 21. Every member of the Association may discontinue voluntarily his membership declaring this in written form to the General Assembly not earlier than a year before leaving.

Chapter Six: General Assembly

Article 22. The General Assembly is the supreme organ of the Association.

Article 23.

1. All members of the Association are invited to participate in the General Assembly, who authorize their representatives for the purpose.

2. The General Assembly is legal if at least two thirds of the representatives of all members are present.

Article 24.

1. The General Assembly is called regularly at least once per year by the Board of Managers.

2. The date, place and agenda of the General Assembly are announced by the Board of Managers at least a month before the event.

Article 25.

1. Extraordinary General Assembly may be summoned by a proposal of two thirds of all members.

2. The proposal, duly signed, is sent to the Board of Managers and obliges it to call Extraordinary General Assembly a month after its handing, with agenda and place of holding pointed in the proposal. Correspondingly defined in Article 24, point 2.

Article 26.

1. The General Assembly is directed by the President.

2. Before opening of the General Assembly the President checks the written credentials of all representatives.

Article 27.

1. Every representative in the General Assembly has the right of one vote.

2. The voting is evident, except in the case of ballot or relieve of duties when voting is secret.

Article 28. The decisions of the General Assembly are taken by a majority of two thirds of the present delegates.

Article 29. General Assembly:

- point 1. Accepts and changes the Statute of the Association.
- point 2. Accepts or approves new members.

- point 3.   Hears out and accepts the report for the activity of the Board of Managers
- point 4.   Conducts an annual audit of the finance activity of the Board of Managers
- point 5.   Approves the budget for the next year.
- point 6.   Accepts the annual plan of the Association.
- point 7.   Works out and accepts short- and long-term programs
- point 8.   Considers and approves decisions for contacts of the Association with juridical and physical persons.
- point 9.   Considers and approves decisions with regard to the aims of the Association.
- point 10. Elects the President and rest members of the Board of Managers.
- point 11. Approves Regulations and other inner documents of the Association.
- point 12. Determines the dimension of the membership fee, order and term of paying.

Article 30. The protocols of the General Assembly are kept permanently in the archives of the Association.

Chapter Seven: Board of Managers

Article 31. In the period between two General Assemblies, the work of the Association is directed by the Board of Managers.

Article 32.

1. The Board of Managers consists of President of the Association, by one representative of every Black Sea country, Co-ordinator and Executive director, who have the decisive vote, as well as of twelve to twenty two scientists with deliberative functions.

2. The members of the Board of Managers are elected by the General Assembly for a term of three years.

Article 33.

The sessions of the Board of Managers are:

- point 1.   Regular,

- point 2.   Extraordinary,
- point 3.   Absence - in the cases of point 1.

Article 34.

1. Regular sessions of the Board of Managers are held once in every four months.

2. The session is legal if two third of the members with suffrage are present.

Members with deliberative functions are also invited to participate.

3. The Board of Managers takes decisions with evident voting by a majority of two thirds of all members with suffrage.

Article 35. Extraordinary session of the Board of Managers is called by a majority of two thirds of all its members with decisive voting according to Article 34, point 2 and point 3, and Article 39 point 2.

Article 36.

1.   The regular sessions of the Board of Managers are held in absence by telephone or television channel.

2.   Thus the taken decisions are sent by fax or in other way by the participants in the session and signed with a note for positive or negative vote.

3.   In sessions of absence, the orders of Article 34, point 2 and point 3, and Article 39, point 2 are applied.

Article 37.

Board of Managers:

- point 1. Calls the members of the Association to a General Assembly,
- point 2. Fulfills the decisions of the General Assembly according to the Statute and local legislation,
- point 3. Rules the Association budget,
- point 4. Prepares the annual finance report for its activity,
- point 5. Reports on the work done before the General Assembly,
- point 6. Accepts new members.

Article 38. Protocols of the sessions of the Board of Managers are permanently kept in the Association records.

Chapter Eight: President

Article 39. The President of the Association:

- point 1.   Presides the General Assembly and sessions of the Board of Managers,
- point 2.   Calls the Board of Managers at sessions, announcing to all members the agenda, way, place and date at least a month earlier,
- point 3.   Represents the Association,
- point 4. Controls the work of the Co-ordinator and Executive Director.

Article 40.

1.   In the absence of the President or when he is not able to execute a definite action, his functions are taken by the Co-ordinator.

2.   The President of the Association, in agreement with the Co-ordinator, may authorize any other member to represent him in a concrete case.

Chapter Nine: Co-Ordinator

Article 41. The Co-ordinator:

- point 1.   Organizes the execution of decisions taken by the General Assembly and Board of Managers,
- point 2.   Controls and develops data Bank "Black Sea",
- point 3.   Reports before the Board of Managers and General Assembly,
- point 4.   Assists in the publicity of unsolved ecological and other problems in the region,
- point 5.   Controls the work of the Executive Director.

Chapter Ten: Executive Director

Article 42. Executive Director:

- point 1.   Organizes and maintains the relations among the Association members,

- point 2. Organizes the preparation and distribution of the Information bulletin of Association,
- point 3. Organizes constant relations with the media,
- point 4. Keeps the records,
- point 6. Keeps the seal,
- point 7. Guarantees for the property of Association,
- point 8. Reports before the Board of Managers and General Assembly.

Chapter Eleven: Property and Finance

Article 43. The Association property consists of movable and immovable possessions, financial resources, author's rights and intellectual property.

Article 44. The Association gathers its material means by:

- point 1. Membership fee,
- point 2. Donations and wills,
- point 3. Subsidies from the country,
- point 4. Any other legal receipts.

Chapter Twelve: Miscellaneous

Article 45.
1. The Association "Clean and Peaceful Black Sea" is permanent.
2. The Association may cease its activity only after a decision of the General Assembly or through the court in cases envisaged within the law.

Article 46. The Association considers its position as a part of the World Association of Towns "Clean and Peaceful Planet EARTH".

Article 47. Symbol of the Association is a stylised cosmic photography of Black Sea region, encircled by two concentric circles and an inscription between them in English and local language "Clean and Peaceful Black Sea".

Article 48. Seal, banner and emblem of the Association contain its symbol.

This Statute is accepted by the Constituent Assembly of the Association, held in the town of Pomorie, Burgas region, Bulgaria on............. 199.. and is subject to registration according to law.

3 October 1990, Sofia, Odessa   19 November 1991, Sofia

DRAFT, October 1990, Pomorie, Odessa, Sofia, Moscow
PROGRAME
of the Association for the period November 1990 - December 1991

1. Publication of Information Bulletin N 1, term 31 November 1990.
2. Preparation for printing the popularized translation in Bulgarian language of the monograph "Applied Ecology of Sea Regions - Black Sea" entitled "Black Sea 1990", term 10 November 1990.
3. Preparation for printing in Russian language of "Black Sea 1990", including materials for Odessa Bay, north-west part of Black Sea, Crimea coast, Sea of Azov, the region of Sochi and the east part of Black Sea, term 15 January 1990.
4. Preparation of contracts for establishment of satellite connection between the towns, E-mail connection and installing of computers with data Bank"Black Sea", 15 January 1990.
5. Organizing the Constituent Assembly of the Association in Pomorie in 1991.
6. Organizing spring, summer and autumn complex expeditions aiming at establishing the ecological responsibility of the separate regions and other sources of pollution.
7. Preparing the annual report of the seminar "Pomorie 91 - Applied Ecology of Black Sea" in November 1991 and its immediate publication.
8. Organizing the General Assembly of the Association in November 1991.
9. Establishment of a joint-stock Black Sea Bank, ship, air-plane and motor transport joint-stock companies and a chain of hotels for recreation and tourism with business centers and yachting harbors
10. Preparing the report of the World Ecological Congress "Brazil 92": "Report -diagnosis on the state of Black Sea with scientific, technological,

management and business recommendations for ecological catastrophe prevention".

Draft Presidential Council
Members of council for a term of 3 years
VALENTIN SIMONENKO, President, Odessa, USSR
NEDELCHO PANDEV, member, Burgas, Bulgaria
        member, Constanza, Rumania
ARKADII SHESTAKOV, member, Sevastopol, Russia
STANISLAV KOSHELEV, member, Mariupol, Ukraine
DJANSUH GUBAZ, member, Suhumi, Abhasia
SULIKO INASHVILI, member, Batumi, Adjaria
NURETIN SYUZEN, member, Istanbul, Turkey

Members of council for a term of 3 years, deliberative vote
STOICHO PANCHEV, Acad., vice-president of B.A.S., Sofia
SIMEON SIMEONOV, PhD, ecologist, VHTI, Burgas
APOSTOL APOSTOLOV, PhD, biologist, IRP, Burgas
ASEN KONSULOV, PhD, biologist, SCI.Secretary, IO B.A.S., Varna
VENELIN VELEV, PhD, geologist, IO B.A.S., Varna
ALEKSANDRU BOLOGA, DSc, oceanologist, Constanza
KARAIVAN GLICHERIE, geologist, Lab.sea geology, Constanza
VALERII MIHAILOV, PhD, director, Odo GOIN, chemist, Odessa
OLEG KUDINSKII, PhD, biologist, Odo INBYUM, Odessa
IGOR ZELINSKII, Acad. AN USSR, Rector OGU, geologist, Odessa
SERGEY ANDRONATI, Acad. AN USSR, chemist, YUZ AN USSR, Odessa
VALERII EREMEEV, DSc., director MGI AN USSR, Sevastopol
YURIY TEREHIN, PhD, MGI AN USSR, Sevastopol
GENADII POLIKARPOV, Acad. AN USSR, biologist, INBYUM, Sevastopol
VLADIMIR MOSKOVKIN, PhD, NITZ ANUSSR, Yalta
IVAN OVCINNIKOV, DSc., oceanologist, YUO IO ANUSSR, Gelendjik
RUBEN KOSYAN, DSc., director YUO IO ANUSSR, Gelendjik
MARAT TSITSKISHVILI, PhD, AN GSSR, Tbilisi
SHALVA DJOASHVILI, DSc., Gruzberegzashtita, AN GSSR, Tbilisi
SERGEY DOROGUNTSOV, Acad. AN USSR, economist, SIPSU, Kiev
VALERII BELYAEV, Acad.. AN USSR, mathematician, Kiev

VICTOR ROMANENKO, Acad. AN USSR, biologist, IGB AN USSR, Kiev
VITALII KEONDJYAN, DSc., physicist, GEOHI AN USSR, Moscow
ALEKSANDER KUDIN, PhD, oceanologist, GEOHI AN USSR, Moscow
MIHAIL VINOGRADOV, Acad.. AN USSR, biologist, IO AN USSR, Moscow
MIHAIL FLINT, PhD, biologist, IO AN USSR, Moscow
VICTOR SAPOZHNIKOV, DSc., chemist, VNIRO, Moscow
VALERII ÊALATSKII, DSc., oceanologist, GOIN, Moscow
NIKOLAI ENIKOLOPOV, Acad.. AN USSR, Director IPM, Moscow, chemist
UMIT UNLUATA, Prof., oceanologist, Inst. for Sea Res., Erdemli
ALTAN AKARA, DSc., oceanologist, Sci. and Tech.Council of Turkey, Ankara
SAMRU UNZL, Prof., biologist, prof. Inst. of Sea Sci. and Tech., Trabzon

Members of council for a term of 5 years
STRACHIMIR MAVRODIEV, PhD, Chief Co-ordinator
VALERIA PERESLAVSKAYA, Executive Director

APPENDIX 1
REPORT-DIAGNOSIS ON THE STATE OF SEA REGION WITH RECOMMENDATIONS FOR MANAGEMENT

1. Management of the ecological state of sea region.
2. Dynamics of the hydrological conditions: observations and modelling.
   2.1 Hydrometeo-regime of region.
   2.2 Modeling and types of circulation.
3. Dynamics of hydrochemical conditions and pollution.
   3.1 Characteristics of the discharged wastewaters - volume, content and regime.
   3.2 Present state of the hydrochemical regime and pollution's in the west part (of entire) Black Sea. Background characteristics.
   3.3 Characteristics of the hydrochemical conditions and pollution's in the region. Exchange atmosphere-sea.

3.4 Modeling the distribution of oil hydrocarbons and some other pollutants in the region.

3.5 Toxicity characteristics of seawater and wastewaters with integral indices.

3.6 Radioactivity of waters, bottom and biological objects.

3.7 Distribution of micro- and macro-elements in water and depositions.

4. Hydrobiological conditions.

4.1 Conditions and after-effects of eutrophication.

4.2 Red tide: type of content, conditions for appearance, after-effects (phenomena of extinction, toxicity and other destruction consequences), possible ways for prevention.

4.3 Classification of pollution sources under the influence of plankton communities.

4.4 Dynamic mapping of anthropogenic transformations of communities around pollution sources.

4.5 Bioluminescence and fluorescence of plankton as an index of anthropogenic influence.

4.6 Content and dynamics of phytoplankton pigments. Photosynthesizing activity.

4.7 Formation of hydrogen sulphide beds in bottom. Dynamics and influence on communities. Destruction after-effects.

4.8 Interaction among communities in coastal (shelf) regions and open sea.

4.9 Accumulation of micro- and macro-elements in biological objects.

4.10 Marine cultures and possibilities for bio-protection of beaches, bio-cleaning of bays. Cultivated mussel and macro-weeds.

4.11 Mnemiopsis leidyi.

5. State of coasts. Geology.

6. State of purification equipments. Science capacity, efficiency, economy.

7. Conclusions.

8. Recommendations.

8.1 Recommendations on management of the ecological situation in the region.

8.2 Recommendations on the purification equipments. Implementation of the newest scientific and technological achievements.

8.3 Recommendations on implementation of ecologically pure technologies.

8.4 Recommendations on protection of coasts.

9. Techeconomical, managing and economical conditions for execution of the recommendations.

APPENDIX 2

Proposal 1

The Association is in search of local and foreign juridical and physical persons as partners in the establishment of international research, business, yachting and tourist center "ECO BLACK SEA" in the town of Pomorie, Burgas district, Bulgaria. The building site is chosen on the seacoast as well as on the coastal part of sea.

Proposal 2

Firm "ECOSY" accepts orders for preparation, short-term (to one year), reports-diagnosis on the ecological state of sea regions with recommendations for naturally conformable management, economical and technological decisions, similar in content to Appendix 1.

Proposal 3

The Association is ready to consider every proposal for cooperation in fundamental and applied ecological investigations, for scientific device construction, for closed technologies of purification equipment's for public, farming industry wastes, for ecologically pure and energy economical technologies in industry, farming and sea industry, for other practical applications of modern science and technology achievements.

Proposal 4

Firm "ECOSY" expects proposals for participation in financing its program of forming a group for fast ecological aid of sea regions.

APPENDIX 3
CLEAN AND PEACEFUL BLACK SEA
WHAT IS GOING ON IN BLACK SEA?

Titles of papers like "Mercy for dolphins", "Anxiety for Black Sea", "Will Black Sea die?", and so on, have been appearing for years in daily newspapers and periodicals of Black Sea countries. A lot of scientists, departmental and academically institutions have made attempts to persuade in most different ways governments and local authorities to take necessary measures for prevention of ecological catastrophe of Black Sea and its bays. Unfortunately, there are many other scientists and their assistants journalists who have reassured the public opinion or such, who spread groundless rumors for self-ignition of sea, catastrophic upraising the level of hydrogen sulphide and even for its flowing in the Mediterranean sea, for the possibility to purify sea from hydrogen sulphide by sulphur production.

The four Black Sea countries accepted the international conventions and regulations for environmental protection. It is hardly surprising for anybody that a big part of the regulations and requirements for naturally conformed exploitation of sea resources is not abided by. Negotiations of signing the Black Sea convention are carrying on for years. International and bilateral programs for investigation of Black Sea and the Danube exist for years too. However, nobody has noticed, except maybe some of the participants, the applications of resources invested in science.

As known, Black Sea is a closed sea, water gathering basin of East and Middle Europe - a region with well developed energy, industry, agriculture, transport and tourism. In the end, all wastes of human activities fall into it. Of course, a part of the substances disintegrate on their way to sea. Lately, however, their volume increased to such an extent that the impact on Black Sea ecosystems began to be noticed considerably.

Also, we have not to forget the undercurrent from the Sea of Marmara through Bosporus and the rainfalls driven by Atlantic and Meditterian cyclones and north-easterly wind that contribute to pollution. The ecological breakdowns in the north-west part of sea, periodic blossoming of plankton, change in quality and quantity of fish catch, decrease of water transparency, constant or temporary bans on bathing along beaches during vacation season, sliming of coastal part of shelf in the west and north-west part of sea, appearance of the new settler Mnemiopsis leidyi (a predatory animal, fallen in Black Sea about 1980 with bails waters of ships arriving from North America), all they illustrate qualitatively this fatal influence on Black Sea ecosystems. There are scientists who call "ecological hysterics" this prediction for the inevitable ecological breakdown of Black Sea. Others believe that the self-cleaning capacity of ecosystems will so change under the influence of anthropogenic pollution that the quality of water and air will remain acceptable for people.

Aim and Obstacles

Our group of scientists, mathematicians, physicists, chemists, biologists and engineers wants to know, not to believe. The impressive development of science and technologies, as well as the powerful computerizing during the last years, is the basis for the numerical study of various non-linear systems - from ecosystems to suitably idealized parts of human community.

We have transformed our way of work from socially based (till 1986; then a program team at the Ministry of Science and Higher Education in 1987, and an international scientific program group "Black Sea" in 1988) into a firm organization. In October 1989 we founded the Firm "ECOSY" for express ecological analyses of sea regions for management of nature-protecting, economical and technological decisions.

Here I must make a note that it is very difficult for us to overcome the scientific, topical, departmental, regional, party and national feudalism.

In the beginning of our work we have considered naively, that if we manage to built an almost adequate model of Black Sea system for predicting the catastrophic development, this will be sufficient to make central and local authorities implement into practice our recommendations for scientifically grounded use.

A deep delusion!

Our next seminar "Pomorie 90 - Applied Ecology of Black Sea" was held from 20 to 30 May 1990. The main purpose of the seminar was to summarize our investigations and write a book "Applied Ecology of Sea Regions - Black Sea 1990". The authors are scientists of about twenty academic and departmental institutes studying Black Sea. The book was printed in the Publishing House of the Ukrainian Academy of Sciences "Scientific

Thought" Kiev in the end of September the same year. Sponsors of the seminar were the Firm "Ecosy", the Regional Council of Burgas. The Bulgarian firm "Microprocessor Systems" ensured a part of the computer technique necessary for insertion, edition and printing the text, tables and pictures.

The book systematizes the results of our work, data for the specially created data bank containing the whole available scientific and historical information on Black Sea, a partial description of the basic sea pollution sources, results from eight specialized expeditions, some results from contactless satellite measurements and report-diagnosis for the state of Burgas Bay with recommendations on commanding the ecological state of May 1989. The book has many defaults and may be mistakes. We shall be sincerely thankful to everybody who may advice or help to fulfill the incompleteness.

Conclusions

1. The sea level raises about millimeter and a half per year in a result of total warming of climate in the north hemisphere, which is a consequence of atmosphere pollution.

2. The number and duration of hurricanes diminishes, as well as the decreasing of wind maximal velocity which decreases the self-cleaning capacity of sea.

3. Since 1986, there has been incessant blossoming of plankton in the north-west and west part of sea from June to August, leading to:

a catastrophic increases of the biomes quantity;

worsening the conditions of bathing, even a danger for health in some regions (Odessa Bay from 1988);

extinction phenomena (dying of bottom organisms due to lack of oxygen and increased water toxicity);

sliming of coastal shelf, which is already an ecological catastrophe for the region because the beautiful sandy bottom with weeds and water transparency change with slime smelling of hydrogen sulphide, opaque water without oxygen to a meter from bottom, toxic.

(Examples: Odessa Bay - gradual disappearance of phillophore weeds till 1987, 1988 - isles of slime, September 1989 - incessant sliming with rate of several tens of centimeters Analogous development with delay of several years is observed in Burgas and Varna Bays);

Remark 1: For the sliming of Burgas Bay and opaque water contributed the discharge of toxic earth mass from excavators work in harbor and ship-building plant, in the region of Sozopol, Pomorie and Stavrova banka, in the summer and autumn of 1989. Since August 1989 new points were defined for discharges of waste earth masses outside the big Burgas and Varna bays. This violation of law is the reason for the lack of tunny in bay in the autumn of 1989. The numerous meetings with Bulgarian and Soviet authorities did not change the situation.

4. The level of hydrogen sulphide is practically the same as in the twenties years of the century, but the thickness of the layer of joint existence of hydrogen sulphide and oxygen has increased, as a consequence of the raised production of oxygen and hydrogen sulphide in the water column;

5. The pollution with oil products of the upper hundred meter layer in open sea exceeds from two to four times the permissible limit concentration of 0.05 ml/l, while in bays and some coastal zones - from tens to hundreds times. This proves that local sources are the main causes for failures in the recreation zones.

6. Pollutions with bionic substances, phenols, surface-active substances, heavy metals, pesticides and other, are similar. For example, in Burgas Bay the pollution with pesticides was average about 50 micrograms per liter in May 1990 (about fifteen years ago in Sweden, the registration of concentrations of pesticides in some river and lake waters of about 1 microgram per liter, caused a stir).

7. In surface microlayer the concentration of polluting substances is from 10 to 1000 times bigger. In coastal regions in close proximity to big towns and rivers, their concentration is four times higher than in open sea. As evaporation depends on surface microlayer properties, this embarrasses a lot the numerical estimations of self-cleaning capacity

8. Four years after the Chernobyl accident, the background radioactivity of water increased from two to three times average. Almost similar is the picture (much more incomplete in isotopic content and concentrations in separate organs) for hydrobionts. According to recent data the background radioactivity has stabilized in its isotopic content and intensity, except Cesium132. The more intensive investigations caused by the Chernobyl

accident and some preliminary results from modeling of the accident show that for the last ten years technogenic radioactivity equal to several Chernobyl's has fallen in sea in unknown ways, i.e. radioactive wastes are discharged cruelly and illegally in Black Sea.

Tracing the sink of technogenic radioactivity after the Chernobyl accident, gives grounds to suppose that the estimation for the water change from bottom layers may be reduced to about 50 years.

Although the ecological perspective for Bleach sea are darker than for Mediterranean, the tourist and health resort conditions on the Black sea' coast are much better. The reason is high bioproductivity of Black Sea, which leads to a short periods for the plankton blooming.

The development of the new settler ctenophore Mnemiopsis leidyi in the last two years turned into a main factor for the catastrophic state of biological communities of pelagic and led to a sharp decrease in the quantity and quality of fish catches in all Black Sea regions, with the exception of turbot. (Turkish and Rumanian fishers catch it, while it has been forbidden for Bulgarian fishers since the spring of 1989.) In 1989 the mass of Mnemiopsis in Black Sea was estimated to 780 million tons wet weight and continued to increase in 1990 as well.

For the present moment nobody may describe numerically and in detail the Black Sea ecosystem death scenario in regions. Obviously, the origination of a new ecosystem will come to birth after two, three years.

The ultimate goal, of course, is clean Black Sea with clean coasts and rivers. This means ecologically clean industrial enterprises, naturally comfortable agriculture, ecologically safe transport, absence of uncontrolled pollution sources and so on. It is necessary to strive for this aim. Much may be done now, before and in parallel with the cardinal change of system.

Our program (accepted in the seminar "Pomorie 88" and defined more accurately in the following seminars) includes:

a) fundamental and applied investigations (complex monitoring with data treatment in real time),

b) elaboration of naturally conformable management, economical and technological decisions for the entire Black Sea ecosystem as well as for separate regions,

c) establishment of a group for quick ecological aid,

d) building purification equipment's and implementation of new ecological safe technologies.

First of all, pollution sources should be identified, localized and controlled, including waters from low powerful local sources. To set in motion their purification in accordance with the ecological requirements: sterilization and detoxification by means of quarz ultraviolet irradiation, hydrogen peroxide, ozonization when necessary, including a phase for anaerobic decay accompanied by biogas production, utilization of bionic elements in biological lakes, cease the discharge of the most toxic polluting substances, including suspended particles.

To stimulate the establishment of an hierarchic system for monitoring and management of Black Sea ecosystem possibly with telescopes in small bays and recreation zones. The program products must guarantee estimation in real time of self-cleaning capacity, in components and volumes of polluting substances of sea as a whole and according to locally chosen hierarchy in regions.

Simultaneously with monitoring, to quicken investigations for estimation of the ecological responsibility of different pollution sources: rivers, coastal towns, Bosporus stream, pollution from atmosphere, pollution's from ships, including military. To supply with the necessary telescopes in the very beginning.

Sea should be divided into zones for recreation, sport, fishing and marine cultures, transport, yield of ores and minerals or bottom types, places for wastes discharge. At that, it is necessary to take into account bottom relief, state of bentos, regime of streams and meteorological conditions in seasons, ventilation of bays and conditions for water mixing and all other hydrophysical, hydrochemical and hydrobiologic factors defining the transfer, utilization and accumulation of polluting substances.

A system for control and management should be worked out for the whole catchment area of Black Sea as well as for its regions in order to take measures at different accidents: from accidents in atomic power stations and big chemical plants to catastrophes with toxic and other loads during shipping, motor, train and plane transport. The first basic part of such a system is the satellite and other contactless monitoring, together with a preliminary established powerful computer net and the corresponding software prepared for work in real time.

To compose groups for quick ecological aid.

To specify norms for permissible limit loads with polluting substances of discharge zones for wastewaters and earth mass according to PLC of separate components and by the summarized and integral indexes. In the region of deep water outfalls of wastes from the big towns (Yalta, Sochi, Istanbul, etc.), the efficient mixing, self-cleaning capacity and time for reaching the surface layers should be estimated.

Particularly big are the uncertainties in forecasting the behavior of parameters in bays and in playing the different scenarios.

The present estimating the value of these investigations (Program agreed in the seminars "Pomorie 88" and "Pomorie 89") is about 60 million levs (roubles) and about 12 million dollars in the course of 3 to 5 years.

ASSOCIATION OF BLACK SEA TOWNS "CLEAN AND PEACEFUL BLACK SEA"

Who will finance these investigations? The analysis of the present economical and political situation, the relations between USSR, Turkey, Bulgaria and Rumania unambiguously show that the governments (and still less the Academies of Sciences) are not able to accept, co-ordinate and bring to the end such a program In our opinion, all interested parties should pay. They are the Black Sea citizens and institutions using Black Sea resources. In short, the Association of Black Sea Towns "Clean and Peaceful Black Sea".

The Association purposes are defined by the name.

The most important concrete tasks are:

1. Financing reports-diagnosis for the ecological state of the local region and implementation into practice of the scientific recommendations by the corresponding authorities within the frames of co-ordinated ecological and economical policy.

2. Establishment of a free economic zone for the efficient use of Black Sea natural resources.

PROJECT

National Inst. of Meteorology and Hydrology, Inst. of Electronics, Iinst. of Nuclear Research and Nuclear Energy, Bulg. Acad. of Sc.
A SYSTEM FOR DECREASE OF THE NEGATIVE AFTER-EFFECTS IN TRANSPORT AND INDUSTRIAL ACCIDENTS IN BURGAS REGION AND THE BIG BURGAS BAY

Duration: 14 months after finance allotment.
STAFF with co-ordinator: Assoc.Prof.,Dr. Strachimir Cht. MAVRODIEV, INRNE, Bulg.Ac.of Sc., Tzarigradsko chossee 72, 1784 Sofia, Bulgaria
tel. 74311 (561, 443), fax 975 36 19.
The complex character of the project puts on the participation of scientists from different Institutes of the Bulg.Ac.of Sc., affiliated institutes and laboratories, Firms, Academically and Departmental Institutes from Russia and Ukraine.
5 December 1991, Sofia

I. SUMMARY

The present achievements in physics, mathematics, chemistry and biology, as well as the impressing development of computational mathematics and computerization give the possibility to solve sufficiently accurately and in real time the complicated hydrodynamic equations for water and air dynamics, and the distribution of different substances in them. The works of

Program group No 435 (in the frames of the international scientific group), in particular the analysis of Burgas Bay pollution's with oil products and the evidence for many times reduction of the official data given in the "Report-diagnosis on the state of Burgas Bay with proposals mastering the ecological situation" (May 1999), are a private example. The program of Firm "ECOSY" (founded on the basis of the international scientific group aiming at overcoming the national, institutional and academically feudalism in performing the Black Sea program) for establishment of a group for quick ecological aid in sea accidents and natural calamities in a sea region is a more serious illustration of the above mentioned.

The NIMH has a long experience in treating the dynamics of atmosphere and water processes, distribution of polluting substances in Burgas valley and Bay, available scientific potential, organization of many expeditions in the region, a stationary net for observation and control of environmental parameters at its disposal, and most important: after reorganization of the B.A.S., there are tendencies and organizational possibilities for applying the results from fundamental and applied investigations. The main purpose of this project is to develop a system for management and decrease of the negative after-effects in:

1.   Industrial accidents leading to release of poisonous substances from the Oil Chemical Combine chimneys near Burgas;
2.   Accidents with tankers carrying or loading oil products in the big Burgas Bay.

II.   DESCRIPTION OF THE PROJECT

The Project includes:

2.1.   Background characteristics and data for the dynamics of atmosphere and Bay

2.1.1. To work out a program product "Burgas" by analogy and on the basis of the program data bank product "Black Sea" and the regular expedition and net measurements and investigations of NIMH in Burgas region, containing the following main blocks:

A.   Background characteristics of atmosphere and Bay pollution;

B. Specification of the actual and potential sources for pollution of the region (transport, industry, agriculture, harbors, airport, etc.) including atmospheric and water transfers;

C. Dynamics of atmosphere and water in bay in several main types of hydrometeorological conditions.

D. Distribution and utilization of different types of substances in atmosphere and water surroundings separately;

E. The interaction atmosphere-sea, river (channel)-sea;

F. Modeling and management in real time in real hydrometeorological conditions of chimney release of a concrete substance;

G. A block for modelling and management in real time in real hydrometeorological conditions of distribution of a concrete oil product (density, content, etc.) at given source parameters;

H. A block with communication information and the ways of its use.

2.1.2 Collecting data for the background characteristics of air, earth and water surroundings, as well as their additional measurement and specification from the informational banks of MOS, Civil Defense, IHPZ-MEA, Ministry of Agriculture and Forests, scientific literature, different departmental records and reports. For this purpose, simultaneous expeditions on land and sea should be organized, using contactless satellite and other measurements. Establishment of a system for monitoring, accident control and announcement on this basis in the frames of 2.1.1.

2.2.   Results on the Subject till Now

2.2.1   Program product data bank "BLACK SEA"

The program product "Black Sea" is in store for filing the natural and anthropogenic characteristics of Black Sea information and its treatment, including mathematical modeling obtained by monitoring.

There are the following possibilities:

1. Formation of sub-bases natural data with a possibility for their fulfillment and formatting, search and selection of information, visualization and editing, if necessary;

2. Express-analysis of chosen data in graphic form: horizontal or vertical distributions of parameters;

3. Calculation of streams and surface water level of a given Black Sea region, according to chosen hydrological data from a sub-base, by using the geostrophic model of circulation. A graphic visualization of the result is possible and also observing the motions in a graphic regime of marker velocity thrown in a given point in the calculated stationary field of the considered region;

4. The distribution picture of anthropogenic pollution and its utilization as a function of source parameters may be observed. The program software of the product is divided functionally into the following four main blocks: - block for choice of the working regime; - block for reading, search, selection and formatting of information; - block for express analysis; - block for modeling.

2.2.2   Past measurements of air surroundings parameters in Burgas.

The investigation of air pollution in towns and industrial zones, in spite of the considerable volume of data piled up till now, is chaotic and does not abide the requirements of the Services for management of situations in anthropogenic accidents and natural calamities.

The selection of points, terms of observation, measured pollutants, methods for treatment of samples and so on, are done without a general concept for the use and information on the measurements.

The optimal system for control and management in accidents takes into account the factors responsible for the pollutants distribution, as well as the distribution structure of pollutants as a result of the interaction among sources, substrate surface, atmosphere dynamics, rainfall. It makes use of recent science achievements, in particular, the numerical modeling of dynamics and diffusion processes in atmosphere in order to get an adequate description of the actual pollution fields in almost real time.

The atmospheric part of the project includes:

1. Complete making of inventory of pollutants and their characteristics: height of chimneys, temperature of released gases, regime of release, dangerous substances and volumes in stock, ways of transport and so on;

2. Climatic characteristic of the region giving information on climate background, where admixtures will spread as well as information on the danger from transfer of additional pollution from neighboring regions;

3. Air pollution characteristic in the region on the basis of long rows of regular data. Thus, the characteristic pollution in the region may be defined on the one hand, and on the other hand it may be bound with the new system for control and management;

4. The nature experiments (including the traced ones) are an essential and irreversible part of the System aiming at control of models, optimization of the points for control and observation, number, accuracy and frequency of the measured parameters, when checking the system real work in practice at civil defense exercise;

5. Application of stochastic and dynamics models in defining the field of harmful substance concentrations in the near ground atmospheric layer over the investigated region, conducted on personal computers in real time;

6. Informational commutation with the existing systems for measurement: national and local hydrometeorological stations and points.

In the region of the Bulgarian Black Sea coast (including Burgas region), the NIMH has a sufficiently dense net of climatic and synoptic stations: one aerological, two ozonic, four stations for control of air pollution, stations for control of air radioactive contamination, three points for control of rainfalls acidity. We should underline that the pollution of atmosphere and waters in Burgas region is also connected with the meteorological conditions and transfer from neighboring regions, for which the NIMH has at disposal a net of stations and rows of data. All meteorological elements are measured in the meteorological stations, as there are one hundred years rows in some of them. The net of stations for air pollution has functioned since 1972 and consisted of four points, situated respectively - two in Varna and two in Burgas. The sample analysis is performed in specialized chemical and radiometric laboratories of the NIMH in Varna and Burgas. Samples are taken in all working days of the year, two or four times per day, for determination of the concentrations of: sulphur dioxide, nitrogen dioxide, hydrogen sulphide, phenol and dust. Samples are taken every day for dry and wet atmospheric deposition for determination of the total beta activity and gamma background. On the basis of these twenty years old data rows, a most general picture may be given for the air background pollution and the radioactive background in the coastal zone as well.

Numerous complex expedition studies have been carried out to obtain a more complete picture of the spatial-time structure of air pollution field and its connection with local climate and meteorological conditions: Burgas 1976; Burgas valley 1990 (in the region of NHK), Varna (1972, 1978), Arkutino 1983; Sozopol 1986 and 1987, Shkorpilovtsi 1977, 1978, 1979, 1982 and 1985, Longoza 1984 and 1985.

The peculiar meteorological conditions in coastal zones and their connection with the distribution of admixtures are investigated by researches of the NIMH and by means of tracer experiments and models for development of the inner boundary layer in sea breeze. The presence of a considerable number of days with breeze circulation along the Bulgarian seacoast favours the complication of the meteorological situation and increasing the content of admixtures in the near ground layer. The established numerical and analytical models for description of the meteorological conditions in breeze circulation can be applied for routine purposes and prognosis of dangerous situations.

Section "Hydrometeorological aspects of air and water pollution" of the NIMH has numerical models at disposal for estimating the fields of concentration and deposition, as a result of remote transfer of pollutants in atmosphere (like SO_2, NOx, NH_3-CH_4, O_3, etc.) working with real meteorological data. For Black Sea, 70% of the river inflow are of the Danube. Bulgarian rivers are considerably smaller, but due to essential disturbances in their quality content, they exert a definite local influence on coastal zone pollution.

A hydrological and hydrochemical net of stations is built for all more significant tributaries in the Black Sea region for watching the loads imported in Black Sea on the Bulgarian part. The outflow and pollution are observed of the following rivers: Batova, Provadiiska, Kamchia, Aitoska, Ropotamo, Veleka and Varna lake including. A net of 21 hydrochemical coastal stations is also built and functioning in the Bulgarian section of Black Sea. The hydrologic net has been in operation since 1940 and the hydrochemical since

1972. For treatment and analysis of samples the National Institute of Meteorology and Hydrology (NIMH) of B.A.S. has two stationary chemical laboratories available in Varna and Burgas, supplied with movable laboratory cars. The following types of pollution are defined: with mineral substances (HCO, SO, CL, Mg, Na, K,..), bionic elements (NO_4, NO, PO_4, F_2), organic substances (K5, dissolved oxygen O_2) and permanganate oxidation (O), specific indices (pH, Mm, T, H). We are able to define the following toxic elements: Pb, Cu, Zn, Cd, As, Se, Cr. The rivers inflow is automatically measured every day, which gives a possibility to calculate the loading of the above mentioned elements introduced in Black sea. NIMH has suitable methods at disposal for determination of loading that Bulgarian rivers introduce in the coastal zone of Black Sea Based on meteorological data.

2.2.3 Lidar investigation of the aerosol pollution field over Burgas valley and NHK.

The aerosol field is simultaneously sounded and mapped over the region by means of two lidars. The horizontal maps at a height of 10-20 m over the surface show the territorial distribution of aerosol field concentration, zones with increased concentration and their evolution. The lidar tactical radius is 5 km in good meteorological visibility. The resolution capacity by space is 50 m. Consequent maps are received every 30 min. After treatment of statistical data from maps obtained for a long period of time to cover different meteorological situations, the average characteristics may be found to serve in the concrete modeling of local pollution. Aerosol maps of the field in vertical plains give information on the vertical distribution of pollution. The influence of the vertical stratification of atmosphere layers on the dynamics and perimeter of distribution of polluting substances is especially significant. The investigation of stratification by lidar proved to be the only possible efficient method. It may sound in particularly important spatial directions and register every minute the changes in the aerosol concentration in 72 consequent atmosphere volumes. Thus, fast running processes in the near ground atmospheric layer can be studied.

The available material equipment consists of a scanning aerosol lidar with laser of copper vapours, sophisticated in recent years and reliable in operation. The lidar is supplied with the necessary computing technique and software.

2.2.4   Express radar information for control and management of ship traffic in Burgas Bay.

2.2.5   Express satellite hydrometeorological and ecological information.

The following parameters and characteristics are measured:

surface temperature;
optical characteristics of surface and atmosphere layers;
characteristics of frontal zones;
velocity of near ground wind;
state of cloudiness;
radiation characteristics (quantity of falling, dispersed and reflected heat);
characteristics of cyclones.

The obtained information may be used for:

perfection of the systems for control on the ecological state and use of natural resources in a given region;
for current and long-time control on the state of environment in the region;
for prognosis of changes in the environmental state;
for discovery and control of critical states of surroundings due to natural calamities and anthropogenic accidents.

CONTENTS OF REPORT- DIAGNOSIS ON THE STATE OF THE BLACK SEA AND THE SEA OF AZOV WITH SCIENTIFIC, MANAGEMENT AND BUSINESS RECOMMENDATIONS FOR MASTERING THE ECOLOGICAL CATASTROPHE

4.4 Contemporary state of the hydrochemical regime and pollution's Background haracteristics.

4.5. Characteristics of the hydrochemical conditions and pollution's Exchange atmosphere-sea.

4.6. Distribution modeling of oil hydrocarbons, Chernobyl radioactivity and some other pollutants in the region.

4.7. Characteristic of toxicity of sea water and wastewaters with integral indices. Express.

4.8. Radioactivity of air, waters, bottom and biote.

4.9. Distribution of macro- and micro-elements in water, biota and depositions.

5. Hydrobiologic conditions.

5.1. Conditions for and after-effects from eutrophication.

5.2. Red tide, type of content, necessary and sufficient conditions for appearance. After effects: phenomena of extinction, toxicity and other destruction after-effects. Possible ways for prevention.

5.3. Classification of pollution sources according to their influence on plankton communities.

5.4. Dynamic mapping of the anthropogenic rearrangements of plankton communities around the sources.

5.5. Bio- and photoluminescence of plankton as indicator of the anthropogenic impact.

5.6. Photosynthesizing activity, content and dynamics of plankton pigments.

5.7. Hydrogen sulphide beds of shelf. Dynamics and influence over communities. Destruction and other after-effects.

5.8. Interaction between communities in the coastal part, shelf and open sea.

5.9. Accumulation of micro- and macro-elements in biological objects.

5.10. Molismology.

5.11. Marine cultures and possibilities for bio-protection of beaches and biopurification of bays. Cultivated black mussel and macro-weeds.

5.12. Mnemiopsis leidyi. Present state, modeling, prognosis.

6.   Sliming of shelf in the north-west part of sea. Causes, dynamics and after-effects.

7.   State of coasts.

7.1. Geology. Ores and minerals. Search and yield of gas, oil and condensate.

7.2. Self-cleaning capacity, earthquake conditions in the west part and technological possibilities of oil companies.

8.   State of purification equipment's

8.1. Science capacity, efficiency, economy.

9.   Conclusions.

9.1. Scientific sufficiency and unsolved tasks.

9.2. Ecology of region. Self-cleaning capacity. Ecological responsibility.

9.3. Transport, communications, tourism, industry and agriculture.

10. Recommendations.

10.1. Management of the ecological situation in the region.

10.2. Closed technologies and purification equipment's

10.3. Implementation of ecologically clean closed technologies - profitable business.

10.4. Protection of coasts.

10.5. Marine cultures, breed fish, introduction of new types and recovery of fish abundance.

10.6. Business and investments: banking, transport, communications, trade and tourism.

11. Management conditions for implementation of the recommendations.

11.1. Convention for Black Sea.

11.2. Association of Black Sea Towns and Regions "Clean and Peaceful Black Sea".

11.3. Advertising the Black Sea region as the most suitable range for demonstration of new world order efficiency, free market and democracy.

May, 1991, Sofia, Pomorie, Moscow

INDEX